Rare Plants of Colorado

Second Edition, 1997

The Colorado
Native Plant Society

Published by Falcon Press® Publishing Co., Inc., Helena, Montana,
and
the Rocky Mountain Nature Association, Estes Park, Colorado,
in cooperation with the Colorado Native Plant Society
Designed by Ann E. Green

Published by Falcon Press® Publishing Co., Inc., Helena, Montana,
and the Rocky Mountain Nature Association, Estes Park, Colorado,
in cooperation with the Colorado Native Plant Society.

Library of Congress Cataloging-in-Publication Data

Rare plants of Colorado / by the Colorado Native Plant Society.—2nd ed.

p. cm.

Includes bibliographical references. (p.) and index.

ISBN 1-56044-529-7 (pbk.)

1. Rare plants—Colorado. 2. Botany—Colorado. I. Colorado Native Plant Society. II. Rocky Mountain Nature Association.

QK86.U6R373 1997

581.68'09788—dc21 96-51430

CIP

Printed in Hong Kong

For extra copies of this book, please check with your local bookstore, or write Falcon Press, P.O. Box 1718, Helena, MT 59624. You may also call toll-free 1-800-582-2665. Books are also available from the Rocky Mountain Nature Association at 1-800-816-7662.

Front Cover Illustration: *Oxybaphus rotundifolius* by Carolyn Crawford

Title Page Photo: William F. Jennings

Introduction Photo: Loraine Yeatts

Table Of Contents

Foreword

The second edition of *Rare Plants of Colorado* is a tangible expression of the Colorado Native Plant Society's purpose, which is to foster ". . . the appreciation and conservation of the Colorado native flora." The Society received many compliments for the first edition, published in 1989. The publication was widely praised as a needed compilation of data, providing impetus for further study of the illustrated taxa. In addition, artist Carolyn Crawford's cover drawing of *Penstemon grahamii* received a prestigious award, the Royal Horticultural Society's Grenfell Silver Medal. We are delighted that another of Carolyn Crawford's drawings graces the cover of this edition.

The enthusiastic reception for the first edition led the Society to commit in 1993 to production of an updated second edition. Since then, the Rare Plant Publication Committee, chaired by Eleanor Von Bargen, has worked to revise and improve the publication. Committee members Janet J. Coles, Miriam L. Denham, William F. Jennings, Susan S. Martin, Velma A. Richards, and Myrna P. Steinkamp have volunteered many hours of effort to make this publication a reality.

Much new information has been added since the first edition of *Rare Plants of Colorado*. We take satisfaction in knowing that some of this information was obtained through research conducted with the assistance of grants from the John Marr Fund, the Colorado Native Plant Society's small grant program for support of research on Colorado native plants.

Throughout the text of this publication you will note comments that various agencies, groups, or individuals have been instrumental in documenting, studying, or protecting our state's rare plants. The Society is grateful to all those whose efforts contribute to the understanding and preservation of Colorado's biological heritage, and we hope this new compilation will be a valuable contribution toward that goal.

— Thomas Ranker, president
Colorado Native Plant Society

Illustration: *Armeria scabra* ssp. *sibirica*

Introduction

The geology, climate, and geography of Colorado have combined to produce an environment in which a highly diverse native flora has evolved. One cannot explore Colorado, from the plains, across the towering peaks of the Rocky Mountains, to the western canyons and high plateaus, without becoming aware of the grand array of plant species inhabiting every possible niche in the landscape. Many of these plants are well known and appreciated; consider, for example, our state flower, the Colorado columbine, *Aquilegia coerulea*. However, along with the familiar plants exist plants seldom seen, and even plants yet to be discovered.

It is the rare plants of Colorado that are the concern of this publication. Although nearly 150 plants are described, we do not wish to convey the impression that these are all of our rare plants. Of the approximately 3,000 plant species native to Colorado, about 10 percent may be considered rare. Included in this publication are the few Colorado plants that are listed as Endangered or Threatened under the provisions of the federal Endangered Species Act. The other rare plants described here are representative of a larger group, the Colorado Plant Species of Special Concern. This list originally was compiled by the Colorado Natural Areas Program (CNAP), a state agency of the Division of Parks and Outdoor Recreation, Department of Natural Resources. Currently, the list is maintained by the Colorado Natural Heritage Program (CNHP), sponsored by The Nature Conservancy. CNAP and CNHP work to identify and provide some form of protection for important natural features of Colorado, including sensitive plant and animal species, special geologic and hydrologic areas, and uncommon or vanishing biological associations. These efforts deserve the fullest support of the people of Colorado, for without such a program much of our natural heritage surely will be lost.

As you will find from the descriptions given for the species, each plant is rare for its own reasons. Some of our rare plants are found only in highly specific habitats and probably never have been widespread. Others were once widespread, but their numbers have diminished or species have been extirpated as a consequence of human activity. There still is much to be learned about most of these rare plants. The Colorado Native Plant Society set an ambitious goal for this publication: to collect the most complete information available for almost 150 of the rarest species of our state and to publish this in a concise, illustrated form that would increase general awareness and understanding of these plants and their habitats. We hope that this publication will stimulate further study of Colorado's rare plants, and of the habitats that appear particularly vulnerable to loss or damage.

Throughout this publication we emphasize that rare plants, in addition to being intrinsically interesting, are markers for rare or special habitats that often result from unusual combinations of environmental factors such as soils, precipitation, temperature extremes, and exposure. Plant rarity is in no way synonymous with "failure," but instead indicates a limited habitat. It is up to us to determine whether the limitation is a natural one or whether the limitation results from the activities of man.

In recent years we have come to understand the importance of biological diversity. We now must realize that loss of diversity is not just a process occurring as the Amazon rain forest is cleared, but that a similar process is occurring right here at home. We must work toward, and encourage our public officials to recognize and support, the preservation of Colorado's biological diversity. In effect, this means we must find ways to preserve biologically significant areas. We must support important private conservation work, such as that of The Nature Conservancy. We must insist that our state and federal land management agencies actively support and manage for the continued existence of all the biological elements within their purview. We must communicate to our legislators how vitally important it is that we legally resolve to maintain the natural heritage of Colorado.

Rare Plants of Colorado is the result of the dedication and expertise of many individuals who value our rich botanical heritage, and who desire to preserve it intact for those to follow. If your awareness, understanding, appreciation, and concern for rare plants is increased through this book, those efforts will have been duly rewarded.

Acknowledgements

The idea for a publication about Colorado's rare plants arose in 1986 during discussions between the Colorado Native Plant Society (CONPS) and the Colorado Natural Areas Program (CNAP), an agency of the State of Colorado. The Colorado Native Plant Society accepted responsibility for such a publication, but acknowledges with gratitude the important and continuing interest and assistance of CNAP. Sincere thanks also go to our publishing partners, Falcon Press and the Rocky Mountain Nature Association (RMNA). RMNA Executive Director C.W. Buchholtz and Publications Coordinator John Gunn were helpful in many ways. We also thank book designer Ann Green for transforming our material and ideas into a finished product.

Principal funding for the first edition of *Rare Plants of Colorado* was provided by the Boettcher Foundation, with additional contributions made by many other organizations and individuals. Funds derived from the first edition, together with support from Falcon Press, enabled the production of this second edition. The Society is grateful for this tangible evidence of concern by so many for the flora of Colorado.

Eleanor Von Bargen, a former president of the Colorado Native Plant Society, served as chair of the Society's Rare Plant Publication Committee during preparation of both the first and second editions of *Rare Plants of Colorado*. Committee members took responsibility for specific aspects of the second edition. William F. Jennings wrote the initial text; Miriam L. Denham coordinated and technically reviewed artwork; Myrna Steinkamp maintained submission records and coordinated photography. Most dot maps were provided by the Colorado Natural Heritage Program; Janet Coles coordinated and verified the maps and prepared those not otherwise available. Miriam L. Denham, Velma A. Richards, and all committee members were responsible for technical editing. Susan S. Martin prepared copy. The Society's sincere thanks go to these committee members for their many hours of work to prepare the second edition.

CONPS members or friends submitted all the photographs and gathered the information used as the basis for both the first and second editions of this publication. Photographs were contributed by John Anderson, Duane Atwood, William L. Baker, Audrey Benedict, James F. Borland, Rick Brune, Janet Coles, Colorado Natural Areas Program (some by unidentified photographers), Carolyn Crawford, Carol Dawson, Jeffrey Dawson, Dale L. Denham, Miriam L. Denham, Hobart N. Dixon, James R. Ferguson, Craig Freeman, Susan M. Galatowitsch, Kenneth D. Heil, William F. Jennings, Jeffrey D. Karron, David Kuntz, Paula Lehr, James H. Locklear, Robert C. Lommasson, Anne Maley, Susan S. Martin, Mesa Verde National Park (M. Colyer and Joe Friedlander), Tamara S. Naumann, Betsy Neely, New England Wildflower Society (Paul M. Brown and others unidentified), Steve L. O'Kane, Jr., J. Scott Peterson, Robert L. Powell, James Ratzloff, Velma A. Richards, Sandy Righter, Rusty Roberts, Peter G. Root, Bobbie Siems, Myrna P. Steinkamp, Bob Thompson, USDA Forest Service (Barry C. Johnston and Dick Roth), Eleanor Von Bargen, William A. Weber, Karen Wiley-Eberle, Dieter H. Wilken, Diane Wilson, M.F. Wojciechowski, and Loraine Yeatts. The Society greatly appreciates their contributions.

Similarly, many artists with a special interest in native plants prepared line drawings for one or both editions of *Rare Plants of Colorado*. The Society thanks Janet Barker, Lynn Bjork, Cynthia Cano, Carolyn Crawford, Genet D'Arcy, Susan M. Galatowitsch, Marjorie C. Leggitt, Jodi McIntosh, Carol Ann Moorhead, Amy Petty, Linda Reeves, Robyn Tierney, Tracy Wager, and Janet Wingate for their outstanding work. Special thanks go to Carolyn Crawford, whose art appears on both covers of this edition as it did the previous one and who has willingly helped in many unseen ways.

For access to their collections, we thank the following herbaria and their curators: The Colorado College, Colorado State University, Denver Botanic Gardens, Gesneriad Gardens (Boulder), Rocky Mountain (University of Wyoming), University of Colorado, University of Northern Colorado, and many others outside our immediate area.

For the second edition, as well as the first, the members of the Rare Plant Publication Committee thank Eleanor Von Bargen for her effective and devoted leadership, without which *Rare Plants of Colorado* would not exist.

Most importantly, we acknowledge with respect the inspiration provided by William A. Weber, professor emeritus and former curator of the herbarium at the University of Colorado. His lifetime of enthusiasm for and unequalled contributions to our knowledge of the native flora of Colorado form the foundation on which this publication rests.

Art Credits

Janet Barker, 90

Lynn Bjork, 20

Cynthia Cano, 57

Carolyn Crawford, Front Cover, 67, 75, Back Cover

Genet D'Arcy, 11, 61

Susan Galatowitsch, 54, 80 upper, 80 lower, 89, 81 right

Jodi McIntosh, 99

Carol Ann Moorhead, 68

Linda Reeves, 64

Robyn Tierney, Foreword, 97

Janet Wingate, 40, 42 upper, 42 lower, 46, 52, 53, 78 upper, 78 lower, 79 upper, 79 lower, 81 left

Alpine Tundra

Alpine tundra, the region above tree limit, is an extreme land — a land of persistent wind, intense sunlight, penetrating cold, and unrestrained openness. The climate of the tundra in the Colorado mountains in some ways resembles that of the Arctic, although seasonal day length is a major difference between the two areas. Rocky Mountain winters above timberline are long and severe, lasting at least 5 months with temperatures almost never rising above freezing. Blizzards can occur any time of year. Summer temperatures rarely exceed 60 degrees Fahrenheit. The growing season may range from 0 to 70 days per year. Although annual precipitation averages 25 inches, desiccating high winds remove snow from windward exposures, leaving areas of dry, almost desert conditions. The windblown snow accumulates in catchment areas nearby, where it may remain up to 11 months of the year. This affords the plants that grow in these catchment areas a ready supply of moisture and protection from the cold temperatures and drying winds.

Vegetation of the Colorado tundra grows primarily upon surfaces that were beyond the reach of the continental Pleistocene glaciation but still heavily influenced by the presence of ice. Permanently frozen ground a few feet below the surface (permafrost) contributed to the creation of unique arctic and alpine surface features. Water expands when it freezes, and the constant expansion and contraction from the freeze-thaw cycles has caused movement that eventually sorts larger rocks from the finer material. This process sometimes produces patterned ground: rocks arranged in polygons, circles, and pavements. In wet areas, ice forms and expands, pushing up and out to cause slow downhill movement of thawed material over permafrost beneath. This process is known as solifluction. Isolated ice masses unevenly elevate the dense, wet turf above them to form irregular columns called frost hummocks.

Tundra plant communities throughout the world look much alike. About half the plants found on the Colorado tundra also grow in the Canadian Arctic and Alaska, as well as in Siberia and Scandinavia. This suggests that during glaciation, refuges for these plants existed not just on the high mountains of the Rockies but also in unglaciated areas within or between ice sheets in the Arctic.

Plant communities usually recognized in the alpine tundra include fellfields, turfs, snow accumulation areas, wet meadows, gopher gardens, and rock areas. Fellfields are windswept, rock-strewn ridgecrests with little water. Here, plants grow as cushions or carpets, an adaptation to reduce water loss. Turfs, gently rolling, lawn-like meadows dominated by sedges, generally are free of winter snow cover. They have the longest growing season and produce the greatest variety of alpine plant life. Plants growing in snow accumulation areas have a short growing season, but they are protected from wind and cold by the insulating effect of snow. Wet meadows occupy depressed areas that freeze in winter and thaw in summer, providing habitat for alpine wetland plants. Where burrowing animals dig into the tundra a few plants have found their niche in the disturbed soil as gopher garden plants. They persist until the turf heals, but may be maintained if a gopher plows it up again. A few plants make use of the accumulation of soil and moisture in rock crevices or under sheltered rock ledges. Some even are found among the unstable rock of talus slopes. Colorado's rare tundra plants occur in several of these communities.

A given rare species often is highly specialized and limited to just one alpine community, or even to a strictly defined habitat within the broader community. For example, the brayas (*Braya humilis* and *Braya glabella)* grow only in moist areas over limestone bedrock. Sea pink, *Armeria scabra* ssp. *sibirica,* occurs only at the base of rocky talus slopes. Alpine fen mustard, *Eutrema penlandii,* is limited to wet meadows.

In spite of its relative inaccessibility, the alpine tundra has been used by humans for a long time. Archaeological evidence dates use back at least 7,000 years. Domestic sheep have grazed the tundra throughout the Rocky Mountains since the early 1900s. Trampling by recreational users has been shown to cause substantial declines in the numbers of plant species present and decreases in plant density and seed production. These uses are of consequence because tundra plants are simultaneously tough and tender. Superbly adapted to survive in a particular microhabitat created by the interaction of microclimate and microtopography, a particular species can endure seemingly forever. However, a tiny alteration — even the removal of a single rock providing shelter, a change in the course of a rivulet, or the compaction of soil by a careless hiker — can destroy the microenvironment so vital to the plant's survival. Because tundra communities are growing at the limits of possible life, small disturbances may result in severe consequences. Tundra ecosystems need to be treated with the utmost care.

Dwarf Columbine, Rocky Mountain Columbine

(Hellebore Family)

Aquilegia saximontana Rydberg

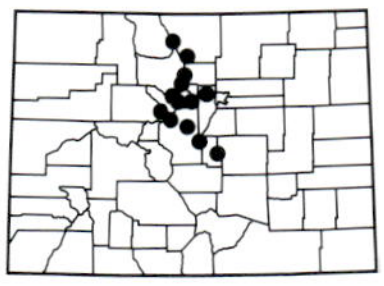

The dwarf columbine, *Aquilegia saximontana,* looks much like a miniature version of the well-known Colorado columbine, *Aquilegia coerulea* James, the state flower and the symbol of the Colorado Native Plant Society.

The dwarf columbine is limited strictly to Colorado. It is found in only a few locations near or above timberline, extending from Larimer to El Paso counties. Typically, it grows in the shelter of large overhanging rocks or in cliff crevices. In comparison to the Colorado columbine's striking long spurs that protrude conspicuously at the back of the flower, the dwarf columbine's spurs are short and incurved, like talons. Dwarf columbine's flowers are blue, but much smaller than those of the Colorado columbine.

Plants grow very slowly in alpine areas, making them particularly susceptible to impact by trampling or off-road vehicles. However, the extremely rocky areas where dwarf columbine grows are not attractive for recreation, so despite its rarity, dwarf columbine seems to be secure at this time.

Von Bargen

Von Bargen

Jennings

Sea Pink (Plumbago Family)

Armeria scabra Pallas ssp. ***sibirica*** (Turcz. ex Bossier) Hylander

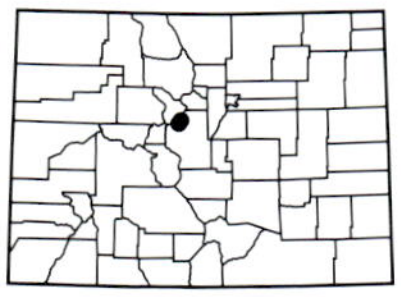

Varieties of sea pink have a widespread and unusual distribution. Occurring not only at high altitude in central and eastern Asia, Scandinavia, and the Colorado Rockies, they also occur on windswept sea cliffs and sandy river banks from Alaska southward to California. Sea pink is particularly conspicuous in the spring at Point Lobos, near Carmel, California. These sea cliff populations give the plant its popular name. The Colorado sea pink, *Armeria scabra* ssp. *sibirica,* is a tundra plant occurring at the base of rocky talus slopes in Park and Summit counties. Sea cliff and tundra plants have much in common, both being assailed constantly by bitter winds. Dwarfed forms are common in each habitat.

From a long, slender taproot and a tuft of basal leaves, several flowering stalks 4 to 12 inches tall arise. Sea pink flowers, which can vary in color from bright pink to magenta, form a globose cluster about an inch in diameter. The flower clusters are quite showy when bobbing around in the summer wind. A closely related sea pink, widely available in the nursery trade, is equally beautiful.

Leadville Milkvetch (Pea Family)

Astragalus molybdenus Barneby

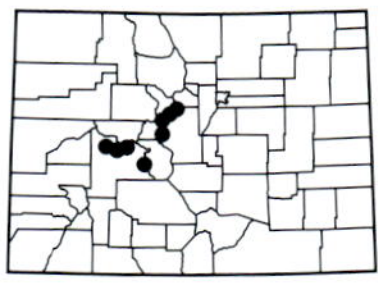

Steinkamp

Dense mats of this inconspicuous milkvetch may be found on dry, gravelly, or rocky tundra underlain by limestone. As is typical of most tundra plants, the Leadville milkvetch is very small, never more than 4 inches tall, usually trailing rather than standing erect. Flowers are violet purple. Leaves and stems have a thick covering of sharp-pointed hairs that are both appressed (lying flat) and ascending, giving the plants a gray look.

In selecting a name for the plant, Rupert Barneby wished to commemorate both the plant's appearance and the locality where it was first collected in 1948, near Leadville, Colorado. *Plumbum* is the Latin word for lead, a gray metal, and Barneby's first choice for a name was *Astragalus plumbeus.* However, that name already had been used for a European milkvetch, so Barneby named his plant for the element molybdenum, also gray and found in the Leadville area.

In Colorado, Leadville milkvetch, *Astragalus molybdenus,* has been found fewer than two dozen times on the high divides near Leadville, Breckenridge, and Fairplay, and in similar habitats in northern and eastern Gunnison County. A population on alpine limestone in Montana is thought by some to be this species. In addition, plants collected from calcareous rocky tundra high on the west slope of the Salt River Range in Lincoln County, Wyoming, look suspiciously like this species. The Wyoming plants, given the name *Astragalus shultziorum,* and the Montana and Colorado plants have been studied intensively, but it is still not clear whether the populations in Montana, Wyoming, and Colorado represent three varieties of a single species or three distinct species. Either way, the plant is rare in Colorado and represents an important source of genetic diversity and an interesting case of extreme disjunction from these close relatives.

D. Denham

Steinkamp

Koenigia (Buckwheat Family)

Koenigia islandica Linnaeus

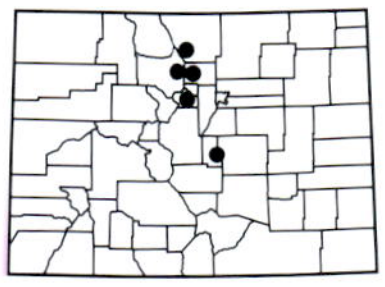

Koenigia islandica is a tiny plant, about ¼ to ½ inch tall. Despite its diminutive size, close observation — preferably with a hand lens — reveals a few pairs of reddish-tinged, oval leaves and several minute flowers that arise at the point of attachment of the leaf to the stem. Amazingly, this is an annual plant. That is, the seeds germinate and the tiny plants produce leaves, flowers, and fruit, all within the short alpine summer. *Koenigia islandica* has been known to botanical science from the days of Linnaeus, the great eighteenth-century botanist who named the species from a specimen found in Iceland.

The occurrence of *Koenigia islandica* in Colorado was unknown until 1953, when William A. Weber was hosting Norwegian botanist Eilif Dahl and others on a field trip to Summit Lake on Mount Evans. Asked what he would show the group that day, Dr. Weber jokingly replied that they surely would see *Koenigia*. Because the plant was known in North America only from the shores of the Arctic Ocean, everyone had a good laugh. The joke was over a few minutes later when Eilif Dahl, on hands and knees, said, "What was that about *Koenigia*?. . . here it is!"

Some years later, Dr. Weber was informed that there was an old specimen of *Koenigia islandica* collected in Colorado and stored at the herbarium in Copenhagen, Denmark. Dr. Weber subsequently had the opportunity to see the specimen, a liter jar of alcohol crammed full of tiny plants of *Koenigia islandica*. Danish botanist Ove Paulsen had collected the plants on Pikes Peak in 1913 while participating in the first International Phytogeographic Excursion. Paulsen didn't realize that the plant was unknown south of the Arctic, and he never published his find.

Martin

SNOW-LOVER (Figwort Family)

Chionophila jamesii Bentham in de Candolle

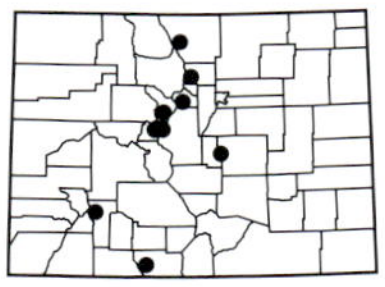

George Bentham was a British botanist with wide-reaching interests. He studied the figwort family extensively and is known to have examined specimens collected in Colorado by Edwin James in 1820 and John Frémont in the 1840s. In 1846, Bentham named the genus *Chionophila*, derived from two Greek words meaning snow and to love.

Snow-lover, *Chionophila jamesii*, is far from common, but it is encountered much more frequently than most of the other rare alpine species discussed in this chapter. We might say that it is the most common of our rare alpine species. It occurs on gravelly, moist slopes above 11,000 feet from southern Wyoming to northern New Mexico. Most are found from Rocky Mountain National Park to Pikes Peak, but snow-lover also is seen in the high mountains of southwestern Colorado. The locations most accessible to wildflower lovers wishing to view this species are along Trail Ridge in Rocky Mountain National Park or on Mount Evans near Denver.

Most alpine plants are small, and snow-lover is no exception. It is usually less than 4 inches in height. Flowers are mainly on one side of the stem (secund). They are yellowish, cream colored, or whitish, and are flattened top-to-bottom when viewed straight on. The flowers darken easily; often they are brown tipped when fresh and black in pressed specimens. In general aspect, snow-lover looks much like a penstemon.

KOTZEBUE GRASS-OF-PARNASSUS (Grass-of-Parnassus Family)

Parnassia kotzebuei Chamisso & Schlechtendal

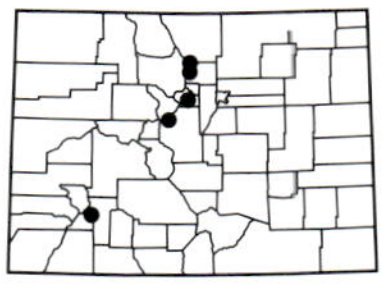

In 1816, when Alaska was under the control of Russia, a scientific expedition financed by Count Nicholas Romanzoff, a Russian minister of state and patron of learning, explored from the Bering Sea to California. Otto von Kotzebue was the expedition's leader, Adelbert Ludwig von Chamisso was the botanist, and Johann Friedrich Eschscholtz was the zoologist. Several plants collected during the expedition subsequently were named for each of these explorers. Kotzebue Sound, the Alaskan town of Kotzebue, and the grass-of-Parnassus were named in honor of the expedition leader.

Like many other rare alpine species, *Parnassia kotzebuei* was known from the Alaskan Arctic for more than 100 years before its discovery in Colorado. Even now, after intensive study of the alpine regions of the Rockies, the few known populations south of the Arctic are scattered in the high alpine tundra of Colorado, Wyoming, Montana, and western Canada.

Despite its common name, this plant isn't a grass. It has been considered to be related to the saxifrages, but results of recent research place it in a separate family. Kotzebue grass-of-Parnassus has small, greenish white, five-petaled flowers. Other species of *Parnassia* in Colorado have a small, leaf-like bract on the stem above a basal rosette of leaves, but the bract is missing in this species. Kotzebue grass-of-Parnassus is an inhabitant of rocky, moist areas, but like so many other tundra plants it may be overlooked easily. So long as the tundra remains mostly remote and undisturbed, this plant probably is in little danger.

Grays Peak Whitlow-grass, Grays Peak Draba (Mustard Family)

Draba grayana (Rydberg) C.L. Hitchcock

Yeatts

Many of Colorado's alpine plants were first seen and collected from the alpine areas at the headwaters of Clear Creek. Compared with most of the rest of Colorado, access to this area was easy 100 years ago because of the gold and silver mining at Central City, Idaho Springs, Georgetown, and Silver Plume. A narrow-gauge railroad ran from Denver directly up Clear Creek to Silver Plume over the famed Georgetown Loop. In the vicinity of Loveland Pass is Grays Peak, a 14,000-foot mountain named for renowned nineteenth-century botanist Asa Gray. It was on Grays Peak that P.A. Rydberg found this draba in 1895. The species was named for both the person and the mountain.

The alpine drabas have been studied in recent years by Robert Price, who reports that Grays Peak whitlow-grass, *Draba grayana*, is restricted to alpine fellfields and scree in northern and central Colorado, mostly east of the Continental Divide. It is found most readily at Summit Lake on Mount Evans, a popular destination that is heavily used in July and August, the peak of the flowering and growth period for tundra plants.

Grays Peak whitlow-grass has yellow flowers and densely tangled, unbranched hairs on its stem, but no hairs on its fruits. The technical feature of hair structure is used by specialists to separate closely related species in the genus *Draba*.

Colorado Divide Whitlow-grass, Twisted-arm Draba (Mustard Family)

Draba streptobrachia Price

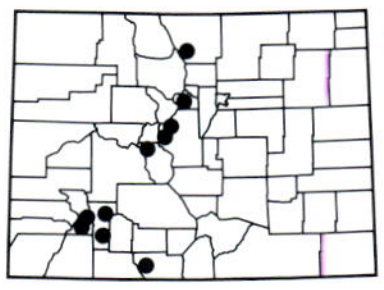

Various species of the genus *Draba* sometimes are called whitlow-grass or whitlow-wort. The name stems from the notion that the plants were a cure for whitlow, an inflamed sore or swelling on a finger or thumb, usually near the fingernail. The technical name for this condition is paronychia. The word whitlow has been in the English language for at least 600 years, so whitlow presumably was a common condition in the Middle Ages. Several plants in other families, including nailwort (genus *Paronychia* in the pink family) and some species of saxifrage, also were regarded by early herbalists as cures for whitlow.

A number of species of *Draba* occur in Colorado. Most are very small alpine plants, rarely over 4 inches tall. Colorado Divide whitlow-grass, *Draba streptobrachia*, is restricted to Colorado. It occurs infrequently in rocky alpine habitats along the Continental Divide, but mainly in the San Juan Mountains. Flowers are about ¼ inch across and have four yellow petals. In the mustard family, the characteristics of the hairs are important for differentiating the species. The hairs of this draba look rather like a piano tuning fork that has been run over by a truck. This deformed appearance of the "arms" of the hairs was the inspiration for an alternative common name, twisted-arm draba, as well as the scientific name.

DWARF BRAYA, ALPINE BRAYA (Mustard Family)

Braya humilis (C.A. Meyer) Robinson var. ***humilis***

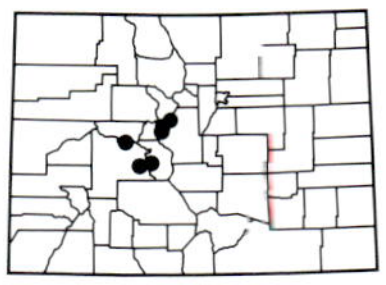

Dwarf braya, *Braya humilis*, was first found in the Altai Mountains of central Asia and named by Carl Anton Meyer, a nineteenth-century botanist who was director of the botanical gardens at St. Petersburg (Leningrad). Since then, several varieties of *Braya humilis* have been found in alpine regions in central Asia, China, Alaska, northwestern Canada, eastern Canada, the northeastern United States, and Greenland The Colorado plants were discovered in 1950. Reed Rollins, an expert on the mustard family, visited Colorado to study the plants and decided they were sufficiently different from the Canadian plants to deserve subspecific status (ssp. *ventosa*). In the mid-1980s, however, James G. Harris of the University of Alberta made a detailed study of the genus *Braya* and concluded that the Colorado plants were not different enough from the typical *Braya humilis* to warrant their own subspecies.

After intensive searching in recent years, populations of *Braya humilis* var. *humilis* have been found in Colorado on limestone outcrops in Park, Summit, Chaffee, Gunnison, and Pitkin counties, but no locations are known elsewhere in the western United States. The nearest populations are 1,000 miles to the north in Alberta, Canada.

As its name suggests, dwarf braya is a small plant, seldom more than 2 inches tall. Leaves, stems, and fruits all bear branched hairs called trichomes. When stems are clustered, as they often are, they produce an attractive display of white flowers with yellow centers, but the flowers are little more than ⅛ inch wide. The deep green leaf rosettes appear to overwinter beneath the snow, allowing the plant to begin growth as soon as the snow melts in spring.

Botanists Betsy Neely and Alan Carpenter, staff members of The Nature Conservancy in Colorado, conducted field studies on dwarf braya that led to a Conservancy purchase of a small preserve on west Hoosier Ridge.

Martin

ARCTIC BRAYA (Mustard Family)

Braya glabella Richardson

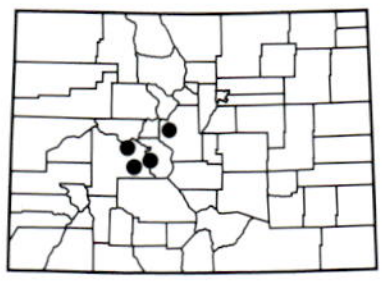

Although *Braya humilis* has been known in Colorado since 1950, *Braya glabella* was not discovered until 1987. Six populations now are known in the high mountains of Gunnison, Chaffee, Park, and Pitkin counties. The species is found in scattered populations across arctic Canada and Alaska, and southward in the high mountains of British Columbia and Alberta. The Colorado populations are 1,000 miles south of the main range of the plant.

The genus *Braya* was poorly understood until it was studied by James G. Harris for his doctoral dissertation at the University of Alberta in the 1980s. Dr. Harris concluded that the specimens collected in Colorado, and some others previously thought to be *Braya humilis*, are in fact *Braya glabella*, a species first described in 1823 from the Arctic. Dr. Harris pointed out that because of isolation and inbreeding there is much variation in species of *Braya*, and although Colorado plants are not identical to plants from northern Canada, they are very similar to the plants from Alberta. This braya once may have had a more general distribution in North America and probably was eliminated from most locations by glaciation.

The flowering stem (scape) of *Braya glabella* has no leaves, in contrast to the leafy stems of *Braya humilis*. The fruits, called siliques in the mustard family, differ as well. Those of *Braya glabella* are short and broad; the fruits of *Braya humilis* are long and narrow. Both plants favor the highly calcareous soils derived from the Leadville Limestone and Manitou Dolomite.

All the known arctic braya populations in Colorado are small, and all occur in areas potentially vulnerable to mining. There is need to seek other populations in likely habitat and to achieve protection for at least some sites.

Martin

GLOBE GILIA (Phlox Family)
Ipomopsis globularis (Brand) W.A. Weber

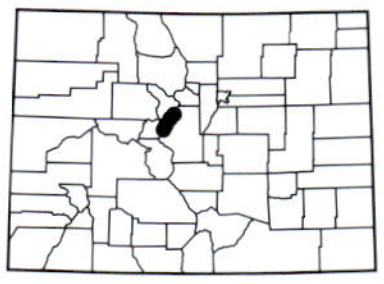

In 1862, C.C. Parry, Elihu Hall, and J.P. Harbour made one of the earliest botanical surveys of the Colorado mountains, collecting in Clear Creek Valley and South Park. Specimens from botanically unknown areas were much in demand in those days, and large sets of collected plants were sold. Thus, this collecting trip was both a commercial enterprise and a scientific endeavor. Evidence indicates that Parry, Hall, and Harbour ascended Georgia Pass (now a four-wheel-drive trail between Jefferson and Breckenridge), where it is likely they found globe gilia. Asa Gray described the plant as a variety of *Gilia spicata*, but in 1907, German botanist August Brand decided the plant deserved specific status and renamed it *Gilia globularis*. William A. Weber of the University of Colorado finally transferred the taxon to the genus *Ipomopsis*.

Although known to botanical science for more than 120 years, *Ipomopsis globularis* remains an elusive plant limited to the Mosquito Range and the Hoosier Ridge area. Globe gilias possess a strong, distinctively sweet fragrance that can be detected many feet away, even before the plants are seen. The creamy white to pale lavender flowers are in a dense, globe-like cluster on a stem about 6 inches tall. Long, silky hairs surround the flowers and give the flower cluster a woolly appearance.

SNOW GRASS, ICEGRASS (Grass Family)
Phippsia algida R. Brown

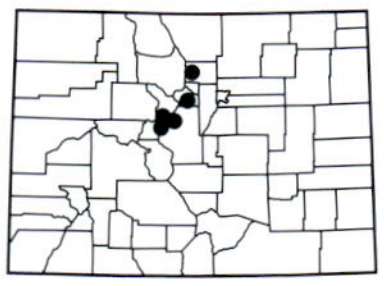

This tiny grass was named from specimens collected in the Canadian Arctic early in the nineteenth century. It was thought to be restricted to the far north for 130 years until William A. Weber of the University of Colorado discovered the species in Colorado in 1952. Since then, a few more populations of *Phippsia algida* have been found in Montana, Wyoming, and on high peaks in the Tenmile, Mosquito, and Front ranges of northern Colorado. Snow grass occurs above 12,000 feet in cold, wet gravel along rivulets fed by melting snowfields. Plants are very low and spreading, generally less than 1 inch tall, and the foliage is somewhat succulent.

Jennings

In 1992, Dr. Weber received a request from a Norwegian botanist for live material. Returning to locations in the Tenmile and Mosquito ranges where the plant was known, Dr. Weber was unable to find any plants. It appeared that drought conditions might have been related to its absence. Finally, plants were found in typical habitat in the Indian Peaks area west of Boulder. Dr. Weber observed that each plant arose separately from a cluster of slender rootlets with no connection to others in the clump. All leaves were green and certainly had been produced during the current growing season. These observations raised the possibility that snow grass might be an annual, not a perennial. The only other known Colorado alpine annual is *Koenigia islandica*, also a rare plant in this state. There are many questions as to how these plants overwinter or survive dry seasons in such a harsh environment. Detailed life cycle studies will be needed to help us understand how these two remarkable alpine plants survive and prosper.

Steinkamp

WEBER SAUSSUREA (Daisy Family)

Saussurea weberi Hultén

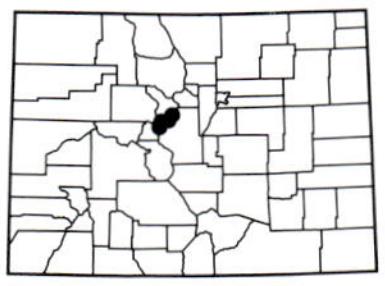

Although first seen in Colorado in 1905, Weber saussurea was not recognized as a distinct species until 1959. It was described by the Swedish botanist Eric Hultén, a specialist in the botany of the far northern regions of the Northern Hemisphere, who named it for long-time University of Colorado botanist William A. Weber. The plant's known range is limited to the high country along the Park-Summit county line, with disjunct occurrences in Wyoming and Montana.

Saussurea weberi is a stout, unbranched plant that grows to about 8 inches in height. The plants, which have long, woolly leaves, produce flowers in rounded clusters of hairy heads. In August, royal purple flowers protrude stiffly from these heads, much like pins from a pincushion. Plants are known not only from nearly barren or disturbed sites, but also from dense tundra rich with grasses, sedges, and small willows.

All the Colorado populations of *Saussurea weberi* are very small and could be vulnerable to alpine grazing or mining. The reasons for the plant's restricted distribution are unknown.

WOOLLY FLEABANE (Daisy Family)

Erigeron lanatus Hooker

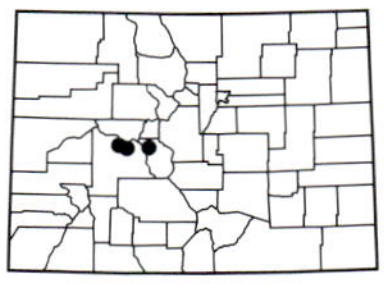

The scientific name for fleabane is *Erigeron*, a combination of the Greek words *eri* and *geron*, meaning early and old man. Although it sounds like a contradiction, the name is ancient, referring to an early blooming, hairy plant. The great Swedish botanist Linnaeus used the name with that meaning in 1753. The species *Erigeron lanatus* has been known to botanical science since the early nineteenth century, when it was described from Canadian plants by Sir William Jackson Hooker, the director of Kew Gardens in London. The specific name *lanatus* means woolly, originating from the Latin word for wool, *lana*. The familiar word lanolin, the oil from wool, has the same root.

Known from Colorado since 1954, woolly fleabane is found on alpine scree slopes above 12,500 feet in Gunnison and Pitkin counties. Joseph Barrell, the person who discovered it in Colorado, noted that the species forms large, compact, beautiful mats more than a foot in diameter. Woolly fleabane blooms in late July and early August. The "wool" of the flower head is white on *Erigeron lanatus*, but purple on *Erigeron humilis*.

Penland Eutrema, Penland Alpine Fen Mustard (Mustard Family)

Eutrema penlandii Rollins

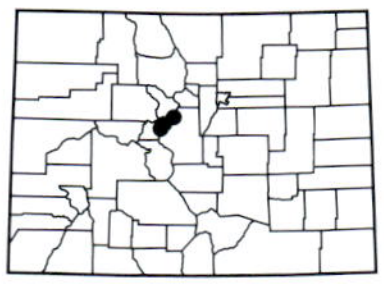

Martin

The Leadville Limestone is a geologic formation of Mississippian age, about 330 million years old. Outcrops occur in the general vicinity of the Colorado city from which it takes its name, as well as near Gilman, Aspen, Tincup, and Marble. In most places, this substrate is highly mineralized and pockmarked by numerous mines and prospects. In the Hoosier Pass area between Fairplay and Breckenridge, the Leadville Limestone underlies the alpine tundra. There, the persistent botanist can find some of the rarest plants in Colorado. The combination of high altitude and high calcium is unique, so a most unusual flora is present. Many unusual species occur only here in Colorado, with their nearest neighbors located in Montana or northwestern Canada. Some species are found here and nowhere else.

Eutrema penlandii, a small member of the mustard family, is a Pleistocene relict whose closest relative is the arctic circumboreal species *Eutrema edwardsii*. All other members of this genus are Asian. Penland eutrema occurs above 12,000 feet on the leeward side of the crest of the Mosquito Range, from Hoosier Pass to Mount Sherman. The plant always is found downslope from persistent snowfields that provide moisture all summer long. Often, the Penland eutrema plants are associated with mosses or other tiny plants, making the eutrema difficult to spot. Most populations are threatened by their proximity to active mines or mining claims. Mining activities that alter surface or ground water supplies can destroy the fragile alpine wetlands required by this species. Penland eutrema is federally listed as Threatened under the Endangered Species Act.

Low Fleabane (Daisy Family)

Erigeron humilis R. Graham

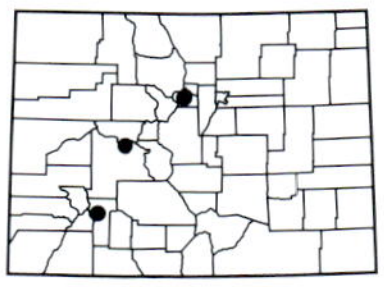

This little fleabane is known from a few scattered sites in the higher latitudes and altitudes of the Northern Hemisphere: Scandinavia, Canada, Alaska, Montana, Wyoming, Utah, and Colorado. The nineteenth-century Scottish botanist Robert Graham, who described the species, grew the plant at Edinburgh. Seeds had been collected in British Columbia in 1827 during one of the early northwestern scientific expeditions. Since its discovery in Colorado in 1954 by ecologist Jean H. Langenheim, populations of *Erigeron humilis* have been found in Gunnison, Pitkin, Hinsdale and Clear Creek counties. At one locality, the low fleabane is associated with *Braya glabella* and *Erigeron lanatus*, also rare alpine plants in Colorado.

From a stem no more than a few inches in height the plant produces a small, typically white-and-yellow, fleabane flower head. The stem is covered with purple hairs. The bracts around the flower head are even more densely hairy, so the flower appears to arise from a ball of purple lint. Because of this feature, the low fleabane presents a striking appearance when viewed from the side.

Grazing and potential mining pose the greatest threats to the low fleabane.

Barrens

The word barrens is loosely defined as any tract of agriculturally unproductive land, but the term is used here to describe sparsely vegetated shale outcrops or shale-derived soils. Areas underlain by the Mancos Shale or Green River Formation tend to develop large expanses of barrens that are prominent features of the plateau region of western Colorado, particularly in the Delta-Montrose-Grand Junction area. Outcrops of the Niobrara Formation along the Eastern Slope, especially in the Pueblo-Cañon City area, tend to develop as barrens. There are even barrens at 12,000 feet elevation near Telluride. Other barrens occur sporadically throughout the state wherever physical and chemical characteristics of the soil or bedrock combine with arid climate to restrict plant growth.

The Mancos Shale and Niobrara Formation were deposited as muds in extensive inland seas of the Cretaceous period, about 65 million to 150 million years ago. The Green River Formation, on the other hand, was deposited in a huge lake during the Tertiary period, about 40 million to 50 million years ago. As the lake expanded and contracted with changes in climate, enormous quantities of muds interspersed with saline minerals were left behind as water evaporated, a process somewhat similar to that occurring today at Utah's Great Salt Lake. Algae and other microorganisms, together with the muds containing them, were transformed into the material known as oil shale. Vast deposits of oil shale potentially exploitable for commercial recovery of oil are found in northwestern Colorado's Piceance Basin. The badlands formations were exposed in a period of extensive erosion that followed regional uplift during mountain-building episodes in Miocene and Pliocene times, 2 million to 20 million years ago.

Climate is an important factor in the development of barren-land plant communities. In Colorado, barrens are best developed in areas that receive less than 12 inches of precipitation per year. The fine texture of clay soils that form from shale prevents the deep penetration or storage of what rain does fall. In marginal areas, a given substrate may support relatively lush and diverse plant communities on cool, moist north-facing slopes, whereas barren-land plant communities occupy warmer, drier, southerly and westerly exposures.

A second important factor in the development of barrens is soil chemistry. Many of the shale formations in Colorado contain high levels of salts, which in excess can be toxic to plants or contribute to drought stress. In other areas, selenium, boron, or other unusual elements in the soil present toxicity problems with which the plants must contend. Many areas are "chalky," rich in gypsum (hydrated calcium sulfate) or other calcium compounds such as limestone (calcium carbonate). Oxygen stress is another problem for barren-land plants. Shale-derived clay soils are composed of very small particles that are easily compacted. Consequently, oxygen exchange and water movement — whether saline or fresh — can be quite restricted. Plant roots require ample oxygen for cell division, water uptake, and growth; thus, oxygen stress curtails plant growth by inhibiting water uptake and root growth.

Botanists use the term endemic to describe plants that are restricted to a particular place or narrowly distributed habitat. The type of endemism exhibited by the plants of barren lands is characterized by extremely local geographic distribution and very specific habitat requirements (often restricted to a single geologic formation or a specific habitat such as steep, eroding slopes). The rare plant sometimes does not appear so rare in a local sense, as many individuals may occur. The Mancos, Green River, and Niobrara formations are hosts to a remarkable number of such endemic rare plants. For instance, Bell twinpod, *Physaria bellii,* is confined to Niobrara Formation outcrops, mostly between Boulder and Fort Collins. The Parachute penstemon, *Penstemon debilis,* is limited to eroding slopes of the Green River Formation.

Some small areas of unusual barrens plants have been recognized as Areas of Critical Environmental Concern and receive management attention from the federal Bureau of Land Management. However, the perceived unaesthetic appearance of barren lands, along with that of their neighbor, the shrublands, has not elicited a groundswell of protectionism. Yet it is on these moonscapes "where nothing grows" that some of the rarest and most endangered plants in Colorado are found. Those of the oil shale country will come under increasing pressure if oil shale mining eventually becomes profitable. In addition, barrens often become popular off-road vehicle playgrounds, with the activity sometimes generally allowed or specific events conducted under permits issued by federal agencies. Detailed botanical inventory and careful consideration of potential effects on the biota should be a minimum requirement before any disruptive activity occurs on barrens, no matter how inconsequential they may appear at first glance.

Martin

Graham Penstemon (Figwort Family)

Penstemon grahamii Keck ex Graham

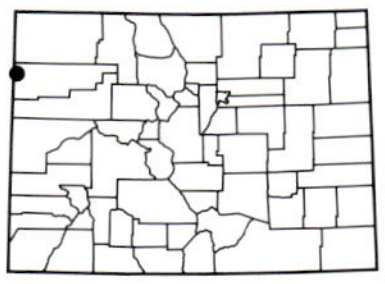

Common names are given to plants by the general public. Such names can both facilitate recognition and cause confusion when the same name is applied to dissimilar plants. On the other hand, rare plants are so infrequently seen by anyone but botanists that true common names are almost never applied. If a common name is wanted, botanists merely translate the scientific name. Graham penstemon is an obvious example, being named after its discoverer, E.H. Graham.

Although known to science since Graham's discovery in 1933, the plant was recorded only a few times in subsequent years. It was always found in eastern Utah on soils derived from the decomposed shales. It was not until 1981 that *Penstemon grahamii* was found in Colorado.

This beautiful little plant evokes comments of joy and delight when seen for the first time. Its special charm led to its choice as the cover subject for the first edition of this publication. Each short, hairy-stemmed plant produces several pale lavender flowers with red stripes. The flowers are proportionally large, appearing much too big for the fragile stem to support. A view directly into the flower reveals what appears to be a face with a wide-open mouth and orange tongue.

Parachute Penstemon (Figwort Family)

Penstemon debilis O'Kane & Anderson

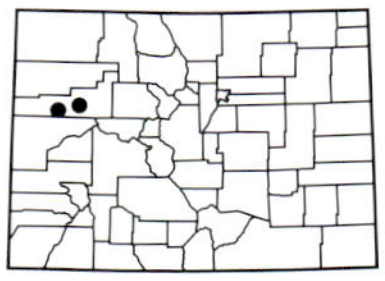

The Piceance Basin of northwestern Colorado is the major resource locality for oil shale in the western United States. Several companies have proposed mining the shale on a grand scale for petroleum products. As a result of such plans, intensive botanical study has occurred there in recent years, leading to the discovery of five species of plants found only on eroding outcrops of the oil shale. While carefully picking their way along the steep slopes of Mount Callahan, botanists Steve O'Kane, Jr., and John Anderson found this unusual penstemon in 1986. Despite repeated searches of likely habitat nearby, the Parachute penstemon, *Penstemon debilis*, currently is known from just two localities.

In the area where *Penstemon debilis* is found, the shale produces a soil of thin rock fragments and clay. Gradually, this soil shifts downhill due to the steepness of the slope. If the penstemon is firmly rooted below the shifting mass it is buried progressively by the shifting shales. The buried stems then elongate downslope until the surface stabilizes enough to allow the development of a tuft of leaves and flowers. Thus, the plant is supremely adapted to its unusual substrate. Leaves are succulent, hairless, and flat blue gray in color; flowers are white to pale lavender.

Gibbens Beardtongue (Figwort Family)

Penstemon gibbensii Dorn

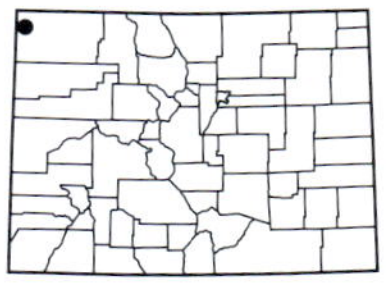

In the early 1970s, Robert Gibbens, a student at the University of Wyoming, collected a penstemon in a remote part of southern Wyoming. Unable to identify the plants, Wyoming botanists sent the specimens to a specialist, but the specimens were lost before they were identified. After 10 years of searching, Robert Dorn, author of floras of both Wyoming and Montana, finally located a small population in 1981. The plants were growing on exposed, barren, sandy clay hills near the Colorado state line. Having live plants and specimens in hand, Dorn decided the beardtongue was distinct and named it for its discoverer.

Colorado botanists independently found the then unidentified species in Moffat County in 1978. Additional sites were found and four populations now are known from a fairly broad area extending from Carbon and Sweetwater counties, Wyoming, through Moffat County, Colorado, into Daggett County, Utah.

Penstemon gibbensii is a small beardtongue, only about 8 inches tall, with several stems arising from the root crown or several plants together forming a clump. Otherwise, the plants are far apart. At one site, Dorn reported 500 plants in an area of 15 acres, an average distance of about 40 feet between plants. Flowers are blue, less than an inch long, and covered with hairs. The hairs on the outside of the flower are short while the hairs inside the flower are longer and broader. Only a few flowers are open at any one time, giving the plants an aspect of poor vigor. Dorn speculated that this may be an indication that the plants once grew under wetter conditions and that the species could be in a long-term decline due to climatic change.

Neely

White River Penstemon (Figwort Family)

Penstemon albifluvis England

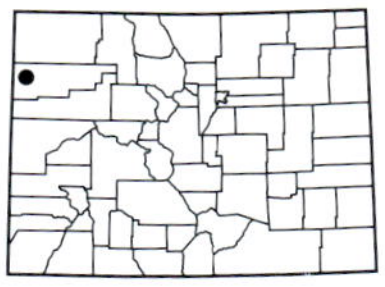

The White River arises in the Flat Tops area of western Colorado and flows westward through Meeker and Rangely before joining the Green River in Utah. The White River penstemon was found growing on the banks of the river and received its scientific name as a result: *albus* means white and *fluvis* means river. Described as *Penstemon albifluvis* in 1982, White River penstemon is sufficiently similar to *Penstemon scariosus* that penstemon specialist Noel Holmgren later wrote that the former could be treated as a variety of the latter. In *Colorado Flora: Western Slope*, William A. Weber comments that this species is very similar to *Penstemon strictus*. Further scientific study apparently is required to determine more precisely the status of the White River penstemon. The White River drainage is part of the oil shale lands of western Colorado and eastern Utah. This penstemon, found on barren shale slopes near the Colorado-Utah state line, shows a few pale lavender flowers with light blue lobes.

Jennings

Mesa Verde Cactus (Cactus Family)

Sclerocactus mesae-verdae (Boissevain & Davidson) L. Benson

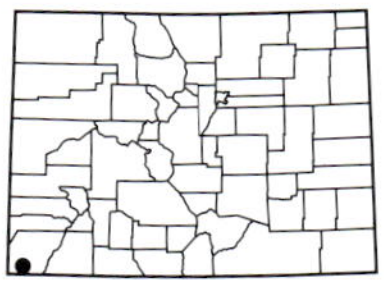

In Montezuma County, Colorado, and San Juan County, New Mexico, a search of the barren, gray clay hills underlain by the Mancos Shale or the Fruitland Formation may reveal a few small plants of Mesa Verde cactus growing exposed to the full strength of the southwestern sun. This is a very harsh environment, with temperatures as high as 110 degrees F, as low as -18 degrees F, and rainfall amounting to only 6 inches annually.

Nevertheless, at the end of April, *Sclerocactus mesae-verdae* produces creamy yellow flowers at the top of the plant. By late June, fruit has matured and seeds fall to the base of the plant. Running water from summertime desert thundershowers seems to be a major factor in carrying seeds away from the parent plants, where they fall into cracks in the adobe soil. Some reports indicate that seeds remain viable for up to 10 years and that prior to germination, the seed must experience numerous freeze-thaw cycles. However, this remains unproven, and the specific requirements for germination are unknown.

Even though the landscape is inhospitable, land use for oil and gas exploration, livestock grazing, powerlines, and pipelines are affecting the plant's habitat. Additionally, nearly all cacti are threatened by commercial and private collectors digging them up for use as house plants. The cactus rarely survives long in cultivation. We can only hope cactus fanciers will learn to be satisfied with photographs of this rare species. Mesa Verde cactus is listed as Threatened under the federal Endangered Species Act and removing live plants is illegal. All known Colorado populations are protected on the Ute Mountain Ute Indian Reservation.

Sun-loving Meadow-rue (Buttercup Family)

Thalictrum heliophilum Wilken & DeMott

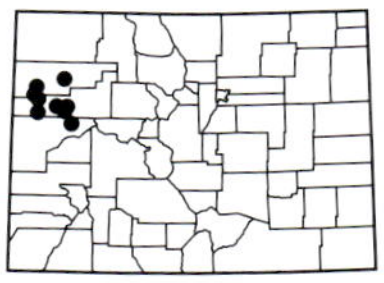

The hydrocarbons known as shale oil are present in the Green River Formation, which comprises huge areas in Colorado, Wyoming, and Utah. When the oil shale is mined, crushed, and heated, the wax-like hydrocarbons are vaporized and can be collected and refined into traditional petroleum products. However, even where oil shale is at its richest, millions of tons of shale must be mined and processed annually to produce enough hydrocarbons for economic feasibility. Because of the large areas that would be mined and additional large areas required for disposal of the spent shale, there is the potential for massive environmental degradation.

Sun-loving meadow-rue, *Thalictrum heliophilum*, is another rare plant discovered during botanical studies conducted in recent years on the oil shale lands. Karen Wiley-Eberle, then a botanist with the Bureau of Land Management, first collected the plant in 1980. It occurs on decomposing shale exposed along the southern and western rims of western Colorado's Piceance Basin and in similar situations in eastern Utah. In contrast to the common meadow-rue, *Thalictrum fendleri*, which occurs in shady aspen woods, *Thalictrum heliophilum* is found on hot, dry shale slopes. Flowering in June and July, sun-loving meadow-rue produces its inconspicuous male and female flowers on separate plants. Leaves of *Thalictrum* species superficially resemble those of columbine (*Aquilegia* sp.), making the plant difficult to distinguish if it is not in flower.

Golden Blazing Star (Loasa Family)

Mentzelia chrysantha Engelmann ex Brandegee

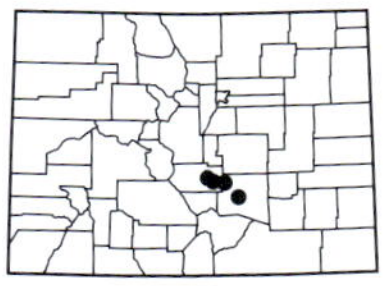

Mentzelia chrysantha was named in 1876 from specimens collected near Cañon City. Twenty years later, E.L. Greene, believing plants he collected in the same vicinity were an undescribed species, described *Mentzelia lutea*. To add to the confusion, both species were recognized by P.A. Rydberg in his 1906 *Flora of Colorado*. The situation was still confused when H.D. Harrington published his *Manual of the Plants of Colorado* in 1954. Dr. Harrington expressed doubts as to the validity of *Mentzelia lutea* and accepted *Mentzelia chrysantha*, but he did not consider them to be names for the same plant. In his 1990 book *Colorado Flora: East Slope*, William A. Weber omitted *Mentzelia chrysantha* and another rare species, *Mentzelia densa*, noting that there was no modern study of this complex.

Our current understanding is that for the plants that occur on limestone barrens east of Cañon City the oldest valid name is *Mentzelia chrysantha*, golden blazing star. The plants called *Mentzelia lutea* are the same species, so this name is a synonym. *Mentzelia densa* occurs on granitic rocks in the Arkansas River Canyon west of Cañon City. The two species are quite distinct in appearance. Golden blazing star is branched but erect, with wide leaves and large, yellow-orange flowers. The Arkansas Canyon stickleaf, *M. densa* (discussed in the Woodlands chapter), forms hemispherical mounds and has narrow leaves and small lemon yellow flowers.

Golden blazing star occurs in habitat similar to that of two other rare limestone barrens species, *Bolophyta tetraneuris* and *Oxybaphus rotundifolius*. The three frequently are found together, and all face the same threats of habitat loss due to mining, grazing, recreation, and urbanization.

Coles

Jennings

Steinkamp

Bell Twinpod, Bell Physaria (Mustard Family)

Physaria bellii Mulligan

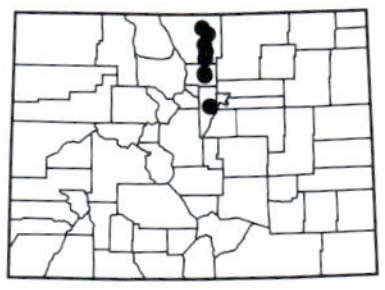

Not far east of the prominent Dakota Hogback between Boulder and Fort Collins is a much smaller ridge where the limestones and calcareous shales of the Niobrara Formation crop out. In two localities, limestones thick enough to justify commercial use for the manufacture of cement have been surface mined. This also is prime habitat for Bell twinpod, which is known only from the Niobrara limestone and shale outcrops in Larimer, Boulder, and Jefferson counties. Although some populations are threatened by the limestone mining, others near Boulder are well protected on city and county open space lands.

About mid-April, shaly outcrops along the road from Boulder to Lyons suddenly are dotted with yellow as the twinpod bursts into bloom. *Physaria bellii* has showy, four-petaled flowers typical of the mustard family. Flowering stems rise both upward and outward from the basal rosette of gray-green leaves, often forming a halo of yellow flowers around the leaves. After flowering, the twinpod develops large, inflated fruits from which the scientific name for the genus is derived; *physa* means bellows or bubble.

Piceance Bladderpod (Mustard Family)

Lesquerella parviflora Rollins

Dudley Bluffs Bladderpod (Mustard Family)

Lesquerella congesta Rollins

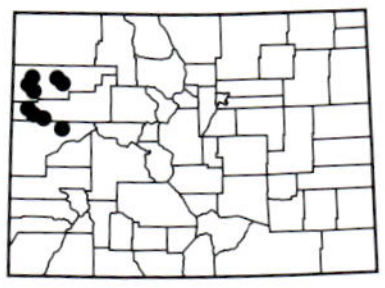

Lesquerella parviflora

Steinkamp

In spite of freeways, state highways, and county roads crisscrossing the countryside, access to vast areas of the West is still slow and difficult. As a result, some areas are very poorly known botanically. This was the case with the Piceance Basin, the area of western Colorado underlain by huge resources of oil shale. Since 1982, five species new to science have been found in the area and it is likely that more will be discovered. *Lesquerella congesta* was found on Dudley Bluffs, a botanically rich area managed by the Bureau of Land Management. Fortunately, Dudley Bluffs is both a designated Area of Critical Environmental Concern and a Colorado Natural Area, providing protection to the plant, which is listed as Threatened under the Endangered Species Act. *Lesquerella parviflora*, another uncommon bladderpod, grows in the same general area, with most populations found along Cathedral Bluffs and Calamity Ridge. Both species grow on nearly barren slopes of weathering oil shale.

Anderson

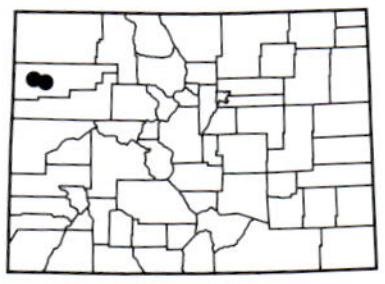

Lesquerella congesta

O'Kane

Dudley Bluffs bladderpod, *Lesquerella congesta*, has a long, thick taproot from which leaves and yellow flowers arise in a congested mass. The plant is only an inch or two in diameter. The scientific name reflects this growth habit. However, it was the trichomes, the specialized leaf hairs, that struck Reed Rollins, who named and described the plant, as most unusual. Some closely related plants can be separated only by highly technical characteristics, and it was the strange hairs of *Lesquerella congesta* that convinced Dr. Rollins that this was an undescribed species.

The trichomes of the Piceance bladderpod, *Lesquerella parviflora*, are equally specialized and unusual. In addition, Dr. Rollins noted considerable diversity in form among plants of this species. He attributed this variability to variation in the microhabitats. Smaller, more compact specimens came from dry, exposed sites; somewhat taller and less compact plants occurred in locations that were more protected. Sparsely branched plants were from shady areas with deep soil.

DeBeque Phacelia (Waterleaf Family)

Phacelia submutica J.T. Howell

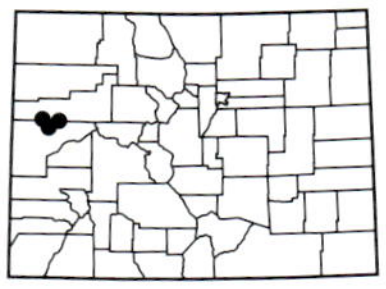

DeBeque phacelia is found at the margins of the Colorado River Valley in Garfield and Mesa counties, in areas of chocolate brown to dark gray clay soils that are too heavy to support most vegetation. The plant appears to be limited to areas of barren soil where there is substantial expansion and contraction. At all known sites there are large cracks in the soil. These soils are derived from the decomposition of carbonaceous strata in the upper and lower (but not the middle) portions of the Wasatch Formation. This formation is slightly older and lies below the oil shale-bearing rocks of the Green River Formation.

From a rosette of elliptic, often reddish leaves arises a stem with minute, cream colored flowers tinged with yellow to purple. The plant is very small; leaves are only ⅝ inch long and flowers are usually less than ⅜ inch across. *Phacelia submutica* blooms in May before the hot summer sun bakes out the little moisture available in these clays.

Jennings

ROUNDLEAF FOUR-O'CLOCK (Four-o'clock Family)

Oxybaphus rotundifolius (Greene) Standley

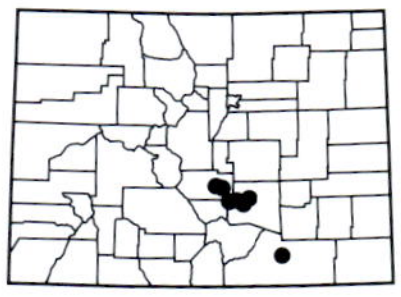

Naumann

The four-o'clock family is characterized by the often tubular, trumpet-like flowers and flower clusters that radiate from a common point like the spokes of a wagon wheel. The cultivated garden four-o'clock is a native of South America. Here in Colorado there are about ten species in the similar genera *Oxybaphus* and *Mirabilis*. One of these species, *Oxybaphus rotundifolius,* roundleaf four-o'clock, is very rare. It grows on limestone outcrops of the Niobrara Formation, particularly between Pueblo and Cañon City. Identified threats to the known populations include mining of limestone for cement production and expansion of housing developments west of Pueblo.

The most common four-o'clock in southern Colorado is *Mirabilis multiflora*, which has very showy 2-inch-long magenta flowers that are seen frequently along roadsides in late summer. The flowers of *Mirabilis multiflora* and *Oxybaphus rotundifolius* are very much alike, except in size. Flowers of roundleaf four-o'clock are smaller, and only reveal their beauty when viewed closely. The leaves of both species are thick and leathery, but the stems of roundleaf four-o'clock are covered with very short, stiff hairs and the stems are less branched. The round leaves, recognized in both the common and species names, are shed early in the growing season and usually are not present when the plant is in bloom. Flowers of roundleaf four-o'clock open very early in the morning and are closed a few hours later. You literally have to get up early to see this plant.

In 1995, Sylvia (Tass) Kelso and her associates at The Colorado College studied roundleaf four-o'clock. This study was timed fortuitously as an unusually wet spring contributed to robust plants and abundant flowering. Surveying areas with the correct combination of substrate and topography, Dr. Kelso's group, together with David Anderson at the Fort Carson army base near Colorado Springs, discovered several new populations of roundleaf four-o'clock. One of these, at the U.S. Army's Piñon Canyon maneuver site in the Purgatoire River drainage, is considerably distant from the other known sites in the Arkansas River Valley, giving hope that even more populations may be found in the future.

PICEANCE TWINPOD (Mustard Family)

Physaria obcordata Rollins

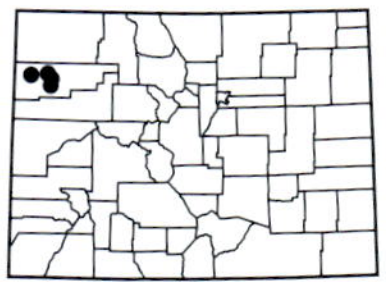

Jennings

While surveying the vegetation of the Piceance Basin in 1982, William L. Baker and Tamara S. Naumann, then of the Colorado Natural Heritage Inventory, found a mustard plant they could not identify as any known species. Reed Rollins of Harvard University, the American authority on the mustard family, named their collection *Physaria obcordata* for the heart-shaped fruit, one of the diagnostic features of the species.

Leaves of the Piceance twinpod form an upright basal rosette. In mid-summer, several stems that are 5 to 8 inches long with yellow flowers project from this leafy base. Piceance twinpod's leaves, stems, and even fruits are covered with specialized branched hairs (trichomes) that are unlike those of any other mustard.

Because of its limited distribution and the threat posed by oil shale mining, Piceance twinpod is classified under the federal Endangered Species Act as Threatened.

CNAP

Pagosa Gilia, Many-flowered Gilia (Phlox Family)

Ipomopsis polyantha (Rydberg) V. Grant

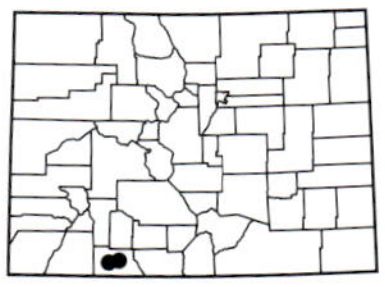

The Mancos Shale weathers into a heavy, gray clay soil. Due to the peculiar crystalline structure of the minerals that make up clay soils, limited water is available to vegetation that may try to grow there. Consequently, vast areas underlain by Mancos Shale form barren badlands. A few plants, however, survive and even thrive in this harsh environment. These plants usually are not abundant, being found only in scattered locations where the conditions are just right. Pagosa gilia, *Ipomopsis polyantha*, is among these hardy Mancos Shale endemics.

Near Pagosa Springs in Archuleta County there are areas in which the Mancos Shale does not weather completely to a homogeneous gray mush, but retains small rock fragments. Here, shrubs such as rabbitbrush (*Chrysothamnus* sp.) not usually found on heavy clays can survive. In the shade of these shrubs, and sometimes where the soil is disturbed at roadside, a few long, white or pink-dotted flowers of Pagosa gilia may be found.

Jennings

Frosty Bladderpod (Mustard Family)

Lesquerella pruinosa Greene

Pagosa Springs, the county seat and largest town in Archuleta County, has become increasingly popular as a resort area. As a result of residential, commercial, ski resort, and airport construction — together with soil disturbance from increased off-road vehicle use around the town — habitat for the frosty bladderpod has been diminished. In addition to the few plants surviving in the population first discovered near Pagosa Springs in 1899, a few other small populations of frosty bladder-pod have been discovered in recent years.

Scattered individuals or clusters of *Lesquerella pruinosa* are found in oak brush, or less commonly in open stands of ponderosa pine, growing in soil derived from the Mancos Shale. The plant does best in shaded areas where organic debris mixes with the clays to form a soil about 6 inches deep over shale bedrock. However, the plant will not tolerate excess shade; it will not grow in the interior of the oak clumps.

This yellow-flowered mustard blooms in May and June. Frosty bladderpod derives both its common and scientific names from the very short, dense hairs that cover the leaves and stems to give the plant a frosted appearance. The diamond-shaped leaves, which narrow abruptly to a long, slender leaf stalk (petiole), are mostly basal and form a rosette against the ground. These features, plus a long taproot that anchors it firmly to the underlying shale, characterize this plant. Another distinctive feature of the species visible in the accompanying photograph is the sigmoid pedicel, the doubly curved stalk of the fruit.

Anderson

Arkansas River Feverfew (Daisy Family)

Bolophyta tetraneuris (Barneby) W.A. Weber

Ligulate Feverfew (Daisy Family)

Bolophyta ligulata (Jones) W.A. Weber

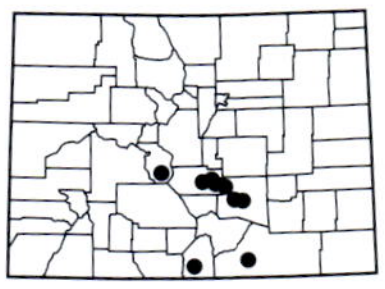

In Fremont and Pueblo counties there are large areas underlain by limestones of Cretaceous age. On barren, raw exposures of bedrock where little else will grow, the Arkansas River feverfew, *Bolophyta tetraneuris,* has managed to find a home. The plant was discovered near the limestone quarry at Portland by Rupert C. Barneby, best known for his work on the genus *Astragalus*. Arkansas River feverfew was known from just two localities until 1983, when intensive searches of the limestone areas revealed nineteen very small populations. Several more have been located subsequently in Chaffee, Costilla, and Las Animas counties in Colorado, and Harding County, New Mexico. Primary threats to the plant's continued survival are limestone quarrying and encroaching urbanization.

Steinkamp

Bolophyta tetraneuris

Martin

Bolophyta ligulata

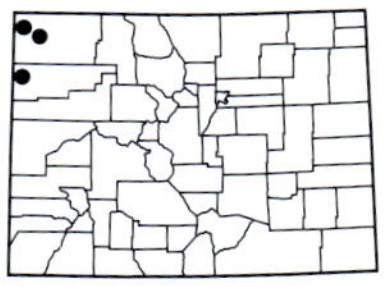

In the remote northwest corner of Colorado, a sister species, ligulate feverfew, *Bolophyta ligulata*, can be found in similar barren situations, but the connection with limestone is lacking. On Raven Ridge, ligulate feverfew grows on oil shale, the rocks bleached white by weathering. In other places in Rio Blanco and Moffat counties and adjacent Utah, the feverfew can be found on barren and sandy clay slopes and flats.

Both species are members of the daisy family (Asteraceae), but that relationship is not obvious at first glance. Like many of the plants of the barrens they are low, mat-forming plants. The thick, branched root is clothed with the remains of dead leaf bases and the few low flower heads are almost hidden by living leaves. The Arkansas River feverfew has white disk flowers only. The ligulate feverfew has both disk and ray flowers, but the rays are very short, about 1/16 inch long. The fruit (achene) of the eastern species has a papery wing. The western species lacks the wing.

Brandegee Wild Buckwheat (Buckwheat Family)

Eriogonum brandegei Rydberg

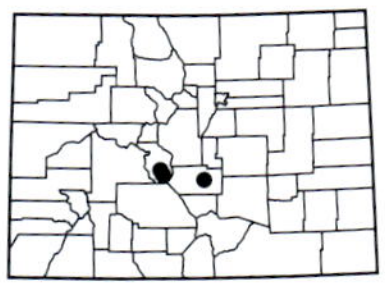

For 95 years after its discovery by pioneer botanist T.S. Brandegee, this attractive wild buckwheat was known only from a small population north of Cañon City. In 1967, James Reveal and Garrett Davidse located a few plants near Salida. Since then, several larger populations have been located around Salida, growing among pinyons and junipers in soils of the Dry Union Formation. Further searches near Cañon City have yielded only one other population, this one growing on the Morrison Formation. The Arkansas River Canyon (Royal Gorge) is cut through igneous rock and divides the two areas of occurrence.

The foliage of *Eriogonum brandegei* is bluish green and has a dense covering of hairs. On the stem and some other plant parts the hairs form loose tufts of "wool," for which the botanical word floccose is used. At the end of a leafless stalk about 8 inches tall there is a globular cluster of minute white flowers with rose highlights. Flowering occurs from late June through mid-August.

The Cañon City population is relatively secure, but the populations near Salida are under pressure from off-road vehicle use on public lands and residential construction on private lands.

Jennings

Steinkamp

Cliffs & Canyons

To understand the plants of the cliff and canyon country one first must understand the geologic processes that created such a landscape. Mountain-building forces deep within the earth lifted the Rocky Mountains and the Colorado Plateau. This process and subsequent erosion exposed colorful sedimentary rock layers that had accumulated through millions of years. The great river systems that arise in the high mountains of Colorado — the Platte, the Arkansas, the Rio Grande, and the Colorado — carved deep canyons into these diverse rocks. The many substrates, in conjunction with constant erosion and variations in solar exposure, have created microhabitats exploited by specialized plants capable of thriving in damp rock fissures.

The principal habitats in cliff and canyon country are moist areas near seeps or springs and dry cracks in the rocks. The wet areas often are under cool overhangs or in alcoves that act like awnings, shading the plants from the baking summer sun. If enough moisture is present, hanging gardens are formed that provide habitat for such rare plants as Eastwood monkeyflower, *Mimulus eastwoodiae;* Kachina daisy, *Erigeron kachinensis*; or giant helleborine, *Epipactis gigantea.* In the dry sites, plants are found sparingly, only where enough soil can accumulate in a crack or where a long taproot can reach available moisture. Some plants of the cliffs and canyons are restricted to certain geologic formations. For example, Black Canyon gilia, *Gilia penstemonoides,* is found only on Precambrian granitic and metamorphic rocks. Rock-loving neoparrya, *Neoparrya lithophila,* is found only on volcanic rocks in southern Colorado.

Native Americans inhabited the canyon country, as evidenced by the cliff dwellings at Mesa Verde where the Cliff Palace milkvetch, *Astragalus deterior,* now is found near the ruins. We do not know the impact — positive or negative — that early Native Americans may have had on such a plant. In general, however, because of their near inaccessibility the rare plants of canyon country have been little affected by the more recent activities of man. These areas usually are unsuitable for livestock grazing, although some canyons are trampled heavily by animals in search of water. Most of the rock formations do not contain valuable minerals and the sites rarely are exposed to wildfires. This does not mean, of course, that the rare cliff and canyon plants are free from danger. Flooding by water storage projects and rock quarrying pose the main threats, but in some areas, even recreational hiking and rock climbing are problems. Recognition of the rarity of several cliff and canyon species has prompted some protective action. For example, near Boulder, the lone Colorado population of black spleenwort, *Asplenium adiantum-nigrum,* has been protected. In Larimer County, the purchase of Phantom Canyon by The Nature Conservancy preserved a population of Larimer aletes, *Aletes humilis*.

Naumann

OWNBEY THISTLE (Daisy Family)

Cirsium ownbeyi S.L. Welsh

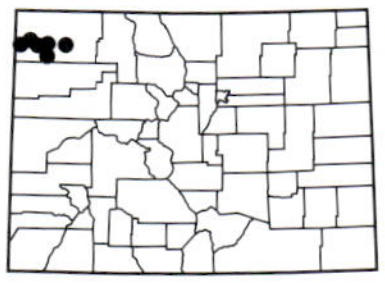

Some familiar introduced species of thistles are serious agricultural pests. However, there are thistles that are a part of the native flora. Ownbey thistle, the rarest of the Colorado native thistles, is found exclusively along the Colorado-Utah state line and in and near Dinosaur National Monument, where several populations occur in canyons along the Yampa River. *Cirsium ownbeyi* was first found in Colorado in 1987.

The plant's typical white or rose pink thistle flowers can be scarcely seen among all the spines that surround the head. The uppermost leaves are reduced almost to spines, and even the deeply cleft lower leaves are spiny. This impressive armature may be the plant's defense against both grazing animals and water loss in the dry canyon country. *Cirsium ownbeyi* was named in honor of Gerald Ownbey, a specialist in thistles who pointed out the distinctive nature of the species.

EASTWOOD MONKEYFLOWER (Figwort Family)

Mimulus eastwoodiae Rydberg

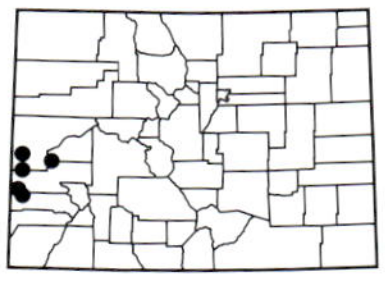

A true hanging garden is an area of luxuriant wildflower growth on a sandstone cliff face. The hanging garden is maintained by a year-round supply of water and a rocky, sandy soil that permits roots to penetrate deeply enough to hold plants in their precarious perch. These oases in the otherwise hot, dry, and rocky canyonlands of western Colorado are home to some strikingly beautiful plants. Because moisture is scarce in the desert, both the habitat and the plants found there are rare and unusual.

In many of the hanging garden communities, particularly at Weeping Rock in Zion National Park in Utah, one can find a bright red monkeyflower, *Mimulus cardinalis*. In a similar habitat in Four Corners country, where Colorado, Utah, Arizona, and New Mexico meet, the persistent botanist may find a few plants of Eastwood monkeyflower, *Mimulus eastwoodiae*. It is also crimson flowered, but lower growing and with more sharply toothed leaves than its relative. Eastwood monkeyflower is stoloniferous, rooting at the nodes (points where leaves, buds, or branches arise). Wherever the root can get a foothold a new plant is formed, even under overhanging rock. The following year, the new plant will be drooping downward, with the flowers hanging below the roots.

Alice Eastwood was the first to collect the plant, but she considered it *Mimulus cardinalis*. P.A. Rydberg recollected the same area, realized the plant was an undescribed species, and in 1913 named it after Eastwood. At most of the known populations in Colorado, Eastwood monkeyflower is associated with cliff columbine, *Aquilegia micrantha*, and the equally rare giant helleborine orchid, *Epipactis gigantea*.

Smith Whitlow-wort, Smith Draba (Mustard Family)

Draba smithii Gilg ex O.E. Schulz

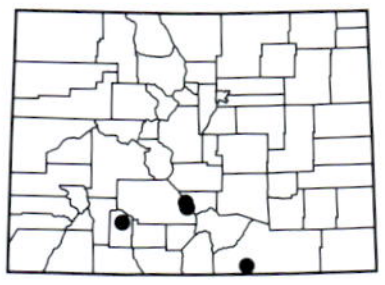

Draba smithii was named by German botanist Ernst Gilg in 1927. Colorado botanist E.C. Smith, for whom the species was named, collected the type specimen in 1891 on a "Mt. Baldy" at 13,100 feet elevation. Within the presently known range of the species, the only location matching the description is Baldy Alto, elevation 13,698 feet and located in the La Garita Wilderness about 10 miles north of Creede. Since 1891, a few more collections have been made at lower elevations in southern Colorado. Populations are known from near Wagon Wheel Gap, the high country of the Sangre de Cristo Range, and the Ratón Mesa-Fishers Peak area above Trinidad. So far as is known, Smith whitlow-wort does not occur in New Mexico, although the Fishers Peak population is less than 12 miles from the state line. However, no in-depth scientific studies have been conducted on this species and little is known other than collection data from herbarium specimens.

The population at Wagon Wheel Gap is most easily accessible, and most who have seen the plant observed it there. Plants occur in protected locations, mostly under overhanging rocks on very steep slopes composed of detritus from the vertical cliff above. At this and some (though not all) other localities, the bedrock is volcanic material. The correlation with igneous rocks appears to be strong enough that new populations should be sought in the San Juan volcanic field west of the San Luis Valley, on volcanic hills in the valley itself, and on the dikes associated with the Spanish Peaks.

Whitlow-worts commonly bloom early in the season, so flowers are not often observed. The flowers in this species are white, but it is the twisted, helical fruits that are distinctive. The plant's leaves, mainly occurring as a basal rosette, appear grayish because they are densely covered with minute, much-branched hairs. Individual hairs can barely be distinguished with the aid of a strong lens. Even the fruits bear fine hairs with branches radiating from a common point (stellate hairs).

Jennings

Yeatts

Jennings

Steinkamp

ROCK-LOVING NEOPARRYA (Parsley Family)

Neoparrya lithophila Mathias

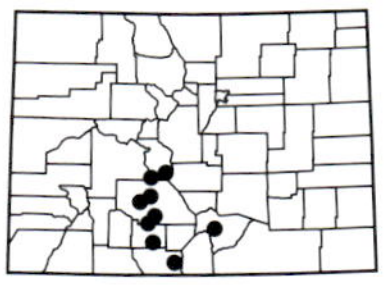

In 1929, after critical examination of specimens collected by C.C. Parry in 1867, Mildred Mathias, a parsley family specialist, decided the plants were so distinctive that they were an undescribed species in a new genus. Many years ensued before *Neoparrya lithophila* was collected again. Its rediscovery in Colorado is one of the better botanical detective stories.

In her original description, Dr. Mathias stated that the Parry specimen had been found at Huerfano Peak in Taos County, New Mexico. By the mid-1950s, after repeated failures at finding the plant in northern New Mexico, she and other workers decided that Parry must have collected the plants elsewhere. William A. Weber of the University of Colorado tried to reconstruct Parry's route. Parry was a member of an expedition studying the feasibility of a railroad route, so documentation did exist as to the location of the party at specific times. But it was Parry's own personal notebooks, preserved at Iowa State University, that proved most helpful as to when and where plant collections were made. The one critical element lacking was the exact date of the collection. Parry's specimen label indicated September, but by early September the party was making its way quickly south to Santa Fe and not collecting plants. Dr. Weber concluded that the only logical place was in the valley below Sangre de Cristo Pass in today's Huerfano County, Colorado. He was right! After a brief search of the area he found rock-loving neoparrya growing in a crevice in a bare rock wall. The date of rediscovery was June 29, 1957 — 90 years after the Parry collection. It was found later that specimens of rock-loving neoparrya had been collected in 1922 in Saguache County by C.E. Taylor. However, Taylor's plants were misidentified and filed away under a different name until the 1980s, when Ronald Hartman of the University of Wyoming recognized them.

Rock-loving neoparrya is a small, yellow-flowered parsley with thick, glossy, leathery leaves. Several small populations have been found in recent years, mostly on volcanic rocks in Chaffee, Saguache, Rio Grande, Conejos, and Huerfano counties.

Hanging-garden Sullivantia (Saxifrage Family)

Sullivantia hapemanii (Coulter & Fisher) Coulter var. ***purpusii*** (Brandegee) Soltis

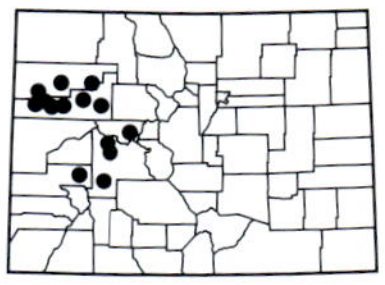

Martin

Wet, dripping rocks seem to be the common factor at the locations, all in western Colorado, where hanging-garden sullivantia has been observed. First seen in the 1890s on cliffs at the bottom of the Black Canyon of the Gunnison, it also is known from seeping or dripping canyon walls in the oil shale country of the Piceance Basin, Hanging Lake in Glenwood Canyon, and above 10,000 feet on quartzite cliffs on Schofield Pass. Perhaps the most interesting locality is in the narrows of East Rifle Creek, where it grows on travertine deposited by water dripping for millennia off limestone cliffs. In this very narrow canyon, ice caves are formed.

Hanging-garden sullivantia originally was described by T.S. Brandegee as *Boykinia purpusii*. Later, it was transferred to the genus *Sullivantia*. Recent research led D.E. Soltis to conclude that it is better treated as a variety of *Sullivantia hapemanii*, a species of northern Wyoming and southern Montana.

Colorado's plants start blooming in June and follow the season to the higher elevations. Sullivantia is about 8 to 12 inches tall with a large number of small white flowers that are about 1/8 inch long in an open panicle. Most of the leaves are at the base. They are five lobed with teeth on the lobes that give them an aspect of a coarse circular saw blade.

Jennings

Canyonlands Lomatium (Parsley Family)

Lomatium latilobum (Rydberg) Mathias

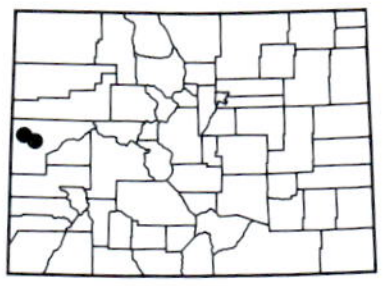

Southwest of Grand Junction in Colorado National Monument, on the steep rocky talus created by disintegration of the majestic red sandstone cliffs and in cracks on slickrock, a cliff-climbing botanist may have the opportunity to see this interesting parsley. Canyonlands lomatium was described by P.A. Rydberg in 1913 from plants found near Moab, Utah. It was first collected in Colorado in 1947 by the well-known botanist Rupert Barneby, who commented, "Neither specimens nor the published descriptions do justice to the size attained by this striking umbellifer, plants of which, composed of many densely cespitose crowns, form great pendent masses on shady ledges of the cliffs."

In addition to the attractive leaf masses, this wide-leaved lomatium has a strong lemony or anise scent. Although *Lomatium latilobum* is known from only a few sites in Mesa and Montrose counties in Colorado and Grand and San Juan counties, Utah, the limited accessibility of its habitat seems to be protecting it from extinction. However, at Arches National Park in Utah some populations are being trampled by heavy recreational use.

Martin

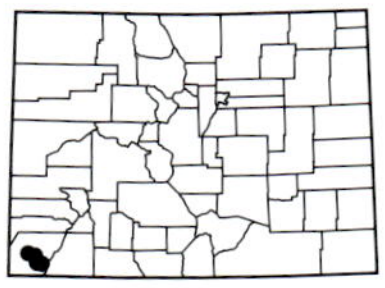

Astragalus deterior

Steinkamp

CLIFF PALACE MILKVETCH (Pea Family)
Astragalus deterior (Barneby) Barneby

NATURITA MILKVETCH (Pea Family)
Astragalus naturitensis Payson

In Mesa Verde National Park, close to some of the most spectacular Anasazi cliff dwellings, grows the Cliff Palace milkvetch. This plant is an excellent example of narrow endemism, defined as requiring for survival a specific habitat that is available in only a small area. Cliff Palace milkvetch is always associated with the white zone of the Upper Cliff House Sandstone, the ledge-forming sandstone at the south rim of Mesa Verde. Found growing on cliffs, in sand-filled depressions on flat or shelving rock, and on loose, sandy talus, the populations do not occur continuously along the rim. Even where the soil and geology are correct, the survival of seedlings is precarious, at best. If June is dry and July is hot, seedling mortality is very high.

Discovered in 1943 by Rupert Barneby, the authority on *Astragalus* in North America, the plant we now know as *Astragalus deterior* was first considered a variety of *Astragalus naturitensis*. A few years later, Dr. Barneby decided it was a distinct species. The small flowers are cream colored with a few purple markings. In the words of Dr. Barneby, Cliff Palace milkvetch ". . .is an inconspicuous little plant." Fortunately, most sites are provided ample protection within the national park, although one population is affected by visitor trampling at a popular viewpoint.

A rare species in its own right, Naturita milkvetch, *Astragalus naturitensis*, occurs mostly along McElmo Creek, the San Miguel and Dolores rivers, and in adjacent Utah. Its habitat consists mainly of sandy depressions in the Dakota and Burro Canyon formations, sandstones similar to the Cliff House Sandstone but somewhat older geologically.

Plants of these two rare milkvetches are similar in size, but there is more purple in the flowers of Naturita milkvetch. Its pods are leathery as opposed to the papery pods of Cliff Palace milkvetch.

CNAP/Coles

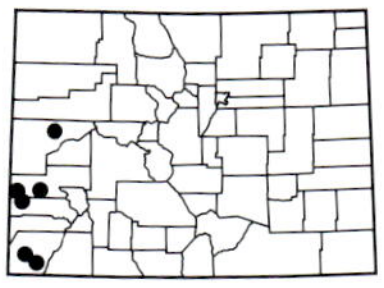

Astragalus naturitensis

KACHINA DAISY, KACHINA FLEABANE (Daisy Family)

Erigeron kachinensis S.L. Welsh & G. Moore

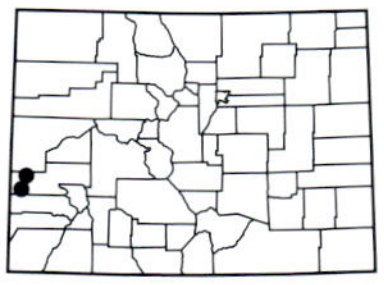

Stanley L. Welsh of Brigham Young University has studied plants of the canyon country for many years, discovering and naming several species previously unknown to science. At Natural Bridges National Monument in Utah, this attractive little fleabane was found growing near Kachina Natural Bridge, one of the primary attractions there. The locality was commemorated in the name *Erigeron kachinensis*.

Kachina daisy is found in wet, seasonally flooded sites and in shallow caves where water seeps through rock cracks to allow development of hanging gardens. Dr. Welsh points out that the Utah cave gardens do not really hang, but they accumulate soil and debris eroded away from the back of the seep to form a sort of cave. As the mass of soil accumulates, it slips forward and downward causing some of the plants to hang precariously until the whole mass falls from the ledge into the canyon and the process starts all over again. Kachina daisy populations may be dense, with individual plants closely spaced. When the plant is in bloom, its small, white, daisy-like flowers with yellow centers add color to the red rock of the sandstone canyons. Unlike most temperate zone plants, this daisy often blooms twice during the year, in spring and again in late summer.

In 1977, kachina daisy was found in Colorado for the first time in the Dolores River Canyon. At present, only two populations are known in the state.

WILKEN FLEABANE (Daisy Family)

Erigeron wilkenii O'Kane

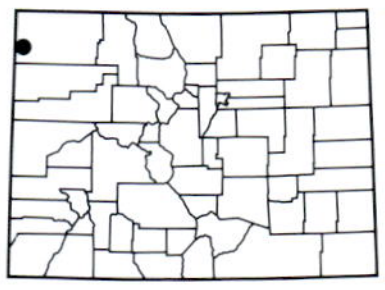

During the first phase of a floristic inventory of Dinosaur National Monument in the summer of 1987, botanists found several species of plants not previously known from Colorado. Along with the other plant collections was this little fleabane, which botanist Steve O'Kane, Jr., described and named for Dr. Dieter Wilken, former professor of botany at Colorado State University.

Wilken fleabane is known only from a small area in Dinosaur National Monument. There, it grows at the base of cliffs in the shade of pinyons and junipers in accumulated duff, often in mats of club-moss in sandy soils derived from Weber Sandstone. In cracks in the canyon walls above, Douglas-fir is found infrequently. It is likely that the fleabane eventually will be found in Utah, less than a mile to the west.

Siems

Growing up to 8 inches tall, Wilken fleabane has a daisy-like white flower with a yellow center. The leaves have short, stiff, sharp-pointed hairs. Dr. O'Kane concluded that the new species was related to the widespread *Erigeron tener*, but like several other recently described species was a narrow endemic existing outside the range of the more common species. The habitat for *Erigeron tener* is usually limestone outcrops, whereas *Erigeron wilkenii* is known only from sandstone.

Jennings

Giant Helleborine (Orchid Family)

Epipactis gigantea Douglas ex Hooker

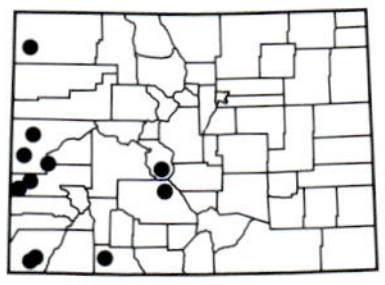

Many people associate orchids with the tropics and find it unbelievable that orchids are found in the desert. But in decomposed sandstone around springs, seeps, and hanging gardens in canyon country, occasionally one will encounter the showy giant helleborine orchid, *Epipactis gigantea*. At higher elevations and in more northern latitudes, it can be found growing in an even more unlikely habitat, near steaming water flowing from hot springs.

The giant helleborine is found occasionally in western North America from central Mexico to southern Canada. It is known in Colorado from only a few locations, all but two on the far Western Slope. Emma Homan Thayer, author of the 1887 book *Wildflowers of the Rocky Mountains*, first reported its existence in this state. The following quotation is from her century-old book:

"On reaching the village, we drove at once up the mountain to see the hot springs that 'will boil an egg in five minutes.' And sure enough, there they were, steaming hot and constantly boiling. One of the attendants at the springs told me of a very rare flower he had found two or three days before, and had in a vase in his room. I expressed a wish to see it, so he brought it to me, and I discovered it to be a wild orchid. The blossom was about the size of a silver dollar and it had the rich brown markings so peculiar to the orchid. I regretted exceedingly that it had been broken from the plant so long that a correct painting could not be given of it. It would have made a valuable addition to my collection."

That population Thayer regretted being unable to paint still exists and is often photographed and enjoyed today.

Black Canyon Gilia (Phlox Family)

Gilia penstemonoides M.E. Jones

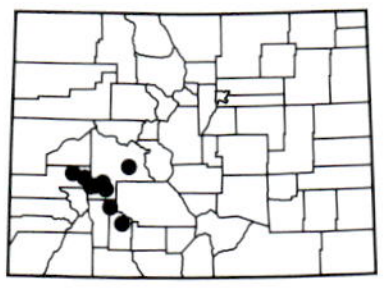

Marcus E. Jones was a mining engineer-turned-botanist who studied western American flora from the 1870s to the 1930s. In 1890, he found an unusual gilia growing in vertical cracks in Precambrian metamorphic rocks in the Black Canyon of the Gunnison and named it *Gilia penstemonoides*. Since that time, this species has been found scattered over Gunnison, Montrose, Ouray, Hinsdale, and Mineral counties in thirteen known populations.

Black Canyon gilia has attractive, trumpet-shaped flowers on 4-inch-long stems which arise from a dense basal rosette of leaves. Flower color in some populations is consistently lavender while in other populations, color tends toward purple. This gilia has an extended flowering period, from early June to late August. Only a few flowers are fresh at any one time. Plants are spaced about 5 to 20 inches apart in the cracks in which they grow.

Some populations of *Gilia penstemonoides* are protected within the Black Canyon of the Gunnison National Monument.

Alcove Death Camas (False Hellebore Family)

Zigadenus vaginatus (Rydberg) Macbride

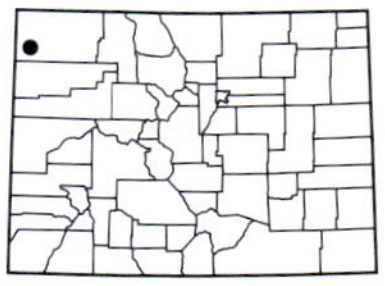

Naumann

The alcove death camas, *Zigadenus vaginatus*, is another rare hanging garden plant endemic to the Colorado River watershed in eastern Utah and western Colorado. From specimens collected at Natural Bridges National Monument, P.A. Rydberg recognized this plant as an undescribed species. In 1912, he named it *Anticlea vaginata*. Later workers included the species in the genus *Zigadenus*. Alcove death camas occurs sparingly in springs, seeps, and hanging gardens in Kane, Grand, and San Juan counties, Utah. It was first collected in Colorado in 1961 at Dinosaur National Monument, but that specimen, stored at Utah State University, was unknown to Colorado botanists. In 1987, during a study of the plants of the monument, Tamara Naumann independently discovered the plant in another area. Several populations were located, scattered in moist locations along the canyons of the Green and Yampa rivers.

Alcove death camas is the giant of this genus, growing from a rootstock almost an inch thick and producing big clumps of leaves and flowering stalks up to 3 feet tall. The plant resembles summer-blooming *Zigadenus elegans*, which is found at higher elevations in the mountains. Alcove death camas is vegetatively larger, has branching inflorescences, and blooms in August and September. Several botanists have pointed out similarities between Colorado plants and those growing in the jungles of Central America, but the exact nature of this relationship is unknown.

Plants in the genus *Zigadenus* (placed by some in *Anticlea*) loosely resemble the onions. The bulbs of wild onions were used by Native Americans and still are eaten by many, but all parts of the death camas plant contain substantial amounts of a toxin and are deadly poisonous, as the name implies. A number of chemically similar toxic compounds have been isolated from other species in the false hellebore family. These compounds, classified as veratrum alkaloids, have been used for more than 100 years in the treatment of certain types of hypertension (high blood pressure).

Neely

Larimer Aletes (Parsley Family)

Aletes humilis Coulter & Rose

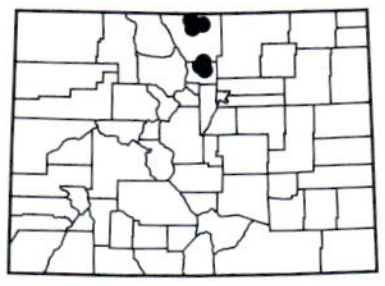

Larimer aletes may appear to be an unimpressive member of the parsley family, but upon careful examination this species, like many rare plants, reveals its own beauty. Growing in cracks and crevices on rocky granite outcrops where it is shaded by ponderosa pine and on cool, near-vertical, north-facing cliffs, Larimer aletes has thick, leathery leaves that help maintain moisture. In the few canyons where it grows, the plant forms conspicuous clumps and mats on rock walls, and when in bloom presents a showy yellow display.

Aletes humilis was discovered in 1898 near Virginia Dale in Larimer County. Nearly a century of looking in Larimer and Boulder counties has yielded only a dozen or so populations, despite the fact that granite outcrops and cliffsides occur nearly everywhere in that part of Colorado. Though rare, Larimer aletes probably is not seriously threatened by normal human activities because the cliffs upon which it grows are generally inaccessible to all but the most determined hikers and climbers. The most serious threat to its survival would be flooding by dam construction. Protection for one population was assured when The Nature Conservancy acquired Phantom Canyon northwest of Fort Collins.

ALCOVE BOG ORCHID (Orchid Family)

Platanthera zothecina (Higgins & Welsh) Kartesz & Ghandi

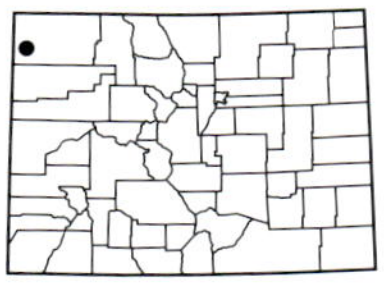

Taxonomy, the science of naming and classifying plants, is an inexact science for two reasons. First, scientists have differing concepts of what constitutes a species. Second, new techniques and more in-depth studies lead to greater knowledge. As a result, complete agreement about names or taxonomic organization of plants can never be expected. Mother Nature also is very uncooperative in this enterprise as she constantly provides plants that defy placement into the neat pigeonholes desired by humans.

North American orchids have been studied extensively during the twentieth century, beginning with P.A. Rydberg in 1901 and continuing with Oakes Ames, Donovan Correll, and Carlyle Luer. All four reached different conclusions regarding the proper taxonomic placement of some orchids. Alcove bog orchid, confined to the hanging gardens in the upper Colorado River watershed in southeastern Utah, northeastern Arizona, and extreme western Colorado, is in a group that has been placed in three different genera. Moreover, the characteristics separating the different species described and named over the last hundred years have been derided as insignificant by some students of the family, yet heralded as of extreme importance by others. Fortunately, Utah and Colorado orchidologists agree that the alcove bog orchid is a distinct species. They do not agree, however, on generic placement. Some Utah botanists believe it belongs in the genus *Habenaria*; some Colorado workers place it in *Limnorchis*. Others argue that it is best considered a *Platanthera*. Until a definitive study is done, we are left with uncertainty regarding the naming of this and similar plants.

Jennings

Known to Utah botanists for many years as being somewhat different from other bog orchids, alcove bog orchid was not named until 1986. Prior to that time it was considered a variant of *Platanthera sparsiflora*. In Colorado, specimens were collected in Dinosaur National Monument as early as 1948, but they were misidentified. It was not until the summer of 1989 when Tamara Naumann, William Jennings, and William Weber saw live plants in Bull Canyon that the identity was confirmed.

As orchids go, *Platanthera zothecina* must be one of the least showy. It has two or three oval leaves near the base of the plant and a few small green flowers along a flowering stalk that is about 6 to 8 inches tall. The most distinctive feature is the long spur that protrudes stiffly out the back of the flower.

Jennings

CNAP/O'Kane

MANCOS MILKVETCH (Pea Family)

Astragalus humillimus Gray ex Brandegee

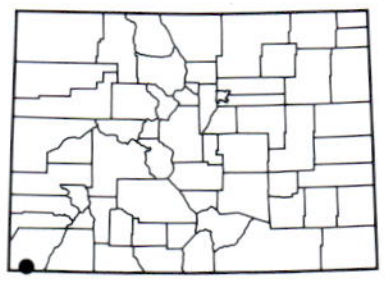

T. S. Brandegee was an engineer-turned-botanist who traveled with an expedition led by Ferdinand V. Hayden into the Four Corners country. He collected specimens of many plants, some of which were very rare and not collected again for many years. In July of 1875, he found a milkvetch "growing upon the sandstone rock of the Mesa Verde, near the edge of Mancos Canyon." Unreported for a century and feared extinct, the plant was rediscovered in San Juan County, New Mexico, in 1980. By 1985, four populations were known, all in New Mexico. After an unsuccessful search in 1985, Colorado botanists Steve O'Kane, Jr., John Anderson, and Betsy Neely returned to the Mancos Canyon area in 1986 and 1987. Finally, small populations of *Astragalus humillimus* were found on the Point Lookout Sandstone near the edge of the canyon, just as Brandegee had found them more than 100 years earlier.

The name *humillimus* means smallest of all, alluding to the low, mat-forming habit of the plant. It grows in sandy soil pockets and stands only an inch or two above them, bearing minute leaflets only about 1/16 inch long. In May and June, Mancos milkvetch produces a pleasing display of flowers that are lavender to purplish with white veins. The most characteristic feature of the plant is the presence of previous years' leaf stalks, which become stiff and spiny when the small leaflets are shed in the fall.

Mancos milkvetch is listed as Endangered under the provisions of the Endangered Species Act.

BLACK SPLEENWORT (Spleenwort Family)

Asplenium adiantum-nigrum Linnaeus

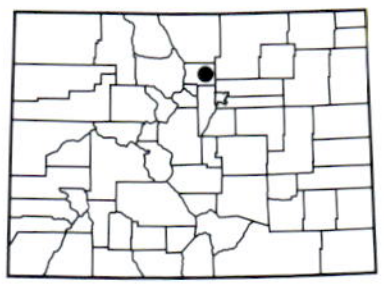

Jennings

In moist seams under an eave-like overhang in a sandstone cliff near Boulder there are a few plants of this small fern, the only known Colorado occurrence. In 1904, Aven Nelson of the University of Wyoming described these plants as a distinct species, *Asplenium andrewsii*. The name honored the discoverer, Boulder nurseryman Darwin Andrews. Other plants of this species were known in Zion National Park, Utah, and near Flagstaff, Arizona. Plants in Mexico, which at that time were considered different species, later were grouped with the Colorado plants.

In the early twentieth century, botanists were not aware that ferns can occur as disjunct populations far from their main range. Fern spores are highly resistant to drying and can be carried great distances by the wind. In many species, only one spore would be sufficient to start a colony. As a result of research in recent years, the western North American occurrences of black spleenwort now are considered as widely disjunct populations of *Asplenium adiantum-nigrum*, a species widespread in Europe and Asia. Even though the fern is no longer considered endemic, the occurrence in Colorado thousands of miles from the main range is most remarkable.

Forests

To many Colorado residents and visitors, forests are the most familiar component of the state's vegetation, although, in fact, forests cover only about 20 percent of the land. Colorado's forests consist mainly of cone-bearing (coniferous) trees, in some settings intermixed with patches of aspen, the deciduous trees well known for their beautiful fall color.

Colorado forests can be differentiated into distinctive life zones beginning at elevations of approximately 5,000 feet, where the lower montane forests border grasslands, shrublands, or woodlands. The forests end at about 11,500 feet, where subalpine forests gradually or abruptly meet alpine tundra.

At the lowest elevations, ponderosa pine, *Pinus ponderosa,* forests are typical. These forests often have an understory composed mainly of grasses, although the species found differ with geography. In other areas, shrubby understories occur. In southern Colorado, for instance, Gambel oak often is present. Elsewhere, bitterbrush, mountain mahogany, wax currant, and squawbush are important forest understory components.

In areas of intermediate elevation, coniferous forests may be more complex, often including two or more tree species in association. Snow may accumulate in these forests to depths of 2 or 3 feet, making the areas cooler and more moist than the ponderosa pine forests. The most common trees are Douglas-fir, white fir, lodgepole pine, and aspen. On north-facing slopes, one often finds closed stands of Douglas-fir associated with such shrubs as jamesia, rocky mountain maple, or ninebark. Open stands of Douglas-fir and ponderosa pine with an understory of grasses occur across the canyon on south-facing slopes.

At higher elevations, Douglas-fir, lodgepole pine, ponderosa pine, and limber pine may form mixed stands. Because Douglas-fir is the most shade tolerant of these conifers it eventually becomes the primary tree of old-growth forests. White fir becomes common south of Colorado Springs. Such white fir forests are primarily southwestern, being widespread in Arizona and New Mexico. In central and northern Colorado, forests of lodgepole pine are common. Where this tree reproduces well, or where there have been recurrent hot fires, lodgepole pine produces a climax or fire-climax community. There is little evidence in these forests that more shade-tolerant trees ever will replace lodgepoles. Understories in these forests typically are very sparse, although there may be a thick layer of old needles.

Above about 9,000 feet, the subalpine coniferous forests include subalpine fir, Engelmann spruce, lodgepole pine, and aspen. Precipitation exceeds 30 inches per year, compared to 18 to 25 inches in ponderosa pine forests and 25 to 30 inches in montane forests. Snow accumulates to depths of more than 3 feet and persists well into June. At the lower elevations in these forests, lodgepole pine and aspen are common, often being replaced gradually by Engelmann spruce or subalpine fir. However, nearly pure stands of lodgepole and aspen may persist for centuries. At the upper elevations of the subalpine forest, the climate is too severe for lodgepole or aspen, and forests consist mainly of subalpine fir and Engelmann spruce. The understories of such forests are highly variable, but they can be rich and luxuriant, especially where moisture is present throughout the growing season. Special ridgetop or upper-slope exposed and windy sites with shallow soils or excessive runoff are environments in which limber or bristlecone pines are found, either in nearly pure stands or sometimes with other subalpine trees.

At timberline, the change from forest to alpine tundra can be abrupt, but a transitional mosaic of trees and tundra also may occur. Here, trees are dwarfed and often become markedly one-sided or flag-form. These wind-sculpted trees form shrub-like islands often called *krummholz,* the very appropriate German word meaning bent tree or crooked wood.

A forest consists of more than just the trees. It is home to communities that include some of the rarest plants in the state. One of the most beautiful rare plants in Colorado, the wood lily, *Lilium philadelphicum,* is disappearing quickly because it is loved too much — populations are decimated by picking or transplanting. Several other plants including white-flowered azalea, *Rhododendron albiflorum*, and trillium, *Trillium ovatum,* are able to exist in Colorado because the forest trees temper the harsh environment, creating conditions approximating those found in cool northwestern forests where these plants are more common.

Except in a few wilderness areas or parks, the forests of Colorado reflect present and past disturbances created by logging, mining, railroads, livestock grazing, skiing, camping, and changes in fire regimes associated with a growing human population. The influence of distant urbanization, particularly in the form of air pollution, is subtle but real. Air pollution may damage forests in ways imperceptible to the eye, but pollution contributes to increases in insect and disease outbreaks and mistletoe parasitism. Thus, even forests that appear to be primeval, pristine, or otherwise minimally influenced by humans nevertheless are affected to some degree.

Jennings

Colorado Yellow Columbine (Hellebore Family)

Aquilegia chrysantha A.Gray var. ***rydbergii*** Munz

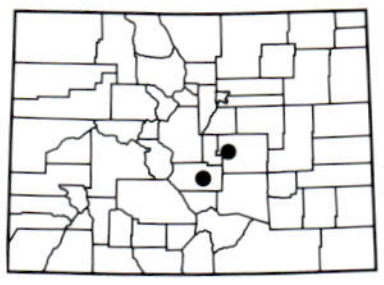

Yellow-flowered columbines have been known from a few forested canyons in southern Colorado since the earliest botanical explorations more than 100 years ago. Other than its color, this columbine looks very much like our state flower. In 1902, P.A. Rydberg decided that the Colorado plants differed enough from the yellow columbines of Arizona and New Mexico that our plants needed a different name. He proposed the name *Aquilegia thalictrifolia* because the leaves reminded him of meadow-rue (*Thalictrum* sp.). However, he did not know that this name already had been used for a European columbine. In botanical science, it is important that each plant has a unique Latin name because duplication of names leads to confusion.

In 1946, Philip A. Munz made a landmark study of the columbines. He determined that the Colorado plants were not sufficiently distinct from the more southerly plants to be a different species. Instead, he proposed that the Colorado plants be recognized only as varietally different from the New Mexico and Arizona plants. Munz was aware of the duplication of names, so he changed the name for our plants to *Aquilegia chrysantha* variety *rydbergii*.

Rydberg commented that the yellow columbine was known from near Cañon City and near Colorado Springs. Forty-four years later, Munz said the same thing. Today, botanists still have not extended that range. It appears that the few populations of Colorado yellow columbine are limited to the El Paso-Custer county area. The nearest populations of the species, but of another variety, are hundreds of miles south near Ruidoso, New Mexico.

Wolf Creek Evening-primrose, Klein Evening-primrose (Evening-primrose Family)

Oenothera kleinii Wagner & Mill

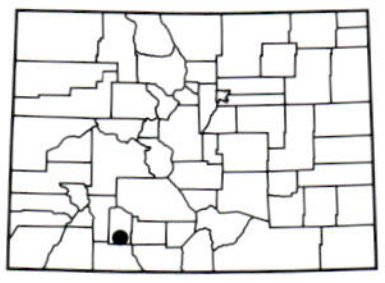

On September 18, 1981, Warren L. Wagner found a unique evening-primrose along U.S. Highway 160 on the west slope of Wolf Creek Pass. Dr. Wagner studied the plant over the winter and concluded that it was an undescribed species, which he decided to name in honor of William M. Klein, director of the Morris Arboretum at the University of Pennsylvania.

Returning to the site in 1982, Dr. Wagner was dismayed to find that a road widening project had eliminated the entire population of *Oenothera kleinii*. No other plants have been found in the vicinity or along the highway to Pagosa Springs. Thus, the species is considered extinct and the original collection is the only known specimen. For purposes of the Endangered Species Act, the plant is considered to be a now-extinct former candidate for listing.

Wolf Creek evening-primrose resembled *Oenothera deltoides*, a species of dry, sandy habitats in the Sonoran and Mojave deserts. It is a most curious association for a mountain forest species. This strange affinity has led to speculation that the roadside plants might not have been native to Wolf Creek Pass but somehow were introduced there. We may never know.

The flowers of *Oenothera kleinii* were about 4 inches across and white with a pale yellow center. Flowers opened at sunset and faded to pink the next day. Leaves were bright green and somewhat hairy with serrate or dentate margins. The plants arose from a thick, fleshy taproot and had a central stem up to a foot tall with branches up to 4 inches long.

Jennings

BROAD-LEAVED TWAYBLADE (Orchid Family)

Listera convallarioides (Swartz) Nuttall

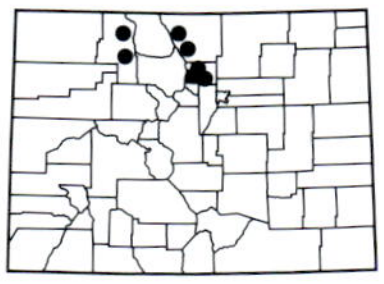

In the moist spray of tiny mountain rivulets usually only a few inches wide, and where there is no substantial riparian habitat like that found along the larger streams, one may find a few plants of this inconspicuous orchid among other streamside greenery. Broad-leaved twayblade, *Listera convallarioides*, often is found with other orchids such as its relative, the heart-leaved twayblade, *Listera cordata*, and bog-orchids (*Limnorchis* spp.) At one locality, twayblade is found with the even more rare *Malaxis brachypoda*.

In Colorado, *Listera convallarioides* grows to about 8 inches. Two rather large, nearly circular opposite leaves at mid-stem account for both the common and scientific names for the plant. Tway is a shortened form of twain, a Middle English word meaning two, that is the basis for the common name, twayblade. The satiny, glossy green color of the leaves reminded the Swedish botanist Olof Swartz of the leaves of lily-of-the-valley (*Convallaria majalis*); hence *convallarioides*, meaning like lily-of-the-valley. Otherwise, the plant doesn't look anything like lily-of-the-valley. Twayblade flowers are green, sometimes with a faint purplish tinge and sometimes so delicate as to be nearly transparent.

Twayblade is found across the Northeast and Canada and southward into the Rockies and the Sierra Nevada. A disjunct population occurs in the Santa Catalina Mountains near Tucson, Arizona. Unknown in Colorado until the early twentieth century, even now there are only a few populations of this orchid known from Boulder, Larimer, and Routt counties. Although it is nowhere common, broad-leaved twayblade has not been afforded any federal protection. All orchids, due to their very specific habitat requirements and overall rarity, are vulnerable to collecting or any significant disturbance in their environment.

WOOD LILY, TIGER LILY (Lily Family)

Lilium philadelphicum Linnaeus

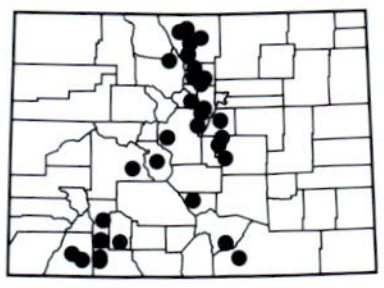

Its cup-size flower of dazzling orange petals with spotted bases makes the wood lily one of Colorado's showiest wildflowers. But this unparalleled beauty is its downfall as it frequently is picked for bouquets or transplanted to urban gardens. Well known to early Colorado botanists, wood lily populations have declined markedly in recent years, and the location of the few remaining stands is usually kept secret. A survey of herbarium specimens has indicated that about two-thirds of the specimens were collected prior to World War II. The lily never has been common in Colorado. When found, it should never be disturbed. Take only photographs.

From its bulb, which was eaten by Native Americans, *Lilium philadelphicum* grows to a height of 1 or 2 feet. Leaves are alternate except for the uppermost, which form a whorl about the stem. There are one to three flowers, each 4 to 5 inches across.

The wood lily is found across most of the northern half of the United States with some occurrences in the Rockies as far south as New Mexico. In Colorado, it is found in moist (but not wet) wooded areas, often in aspen groves or meadows within wooded areas, usually between 7,000 and 9,000 feet elevation. It requires some shade, but does best in relatively open areas.

Martin

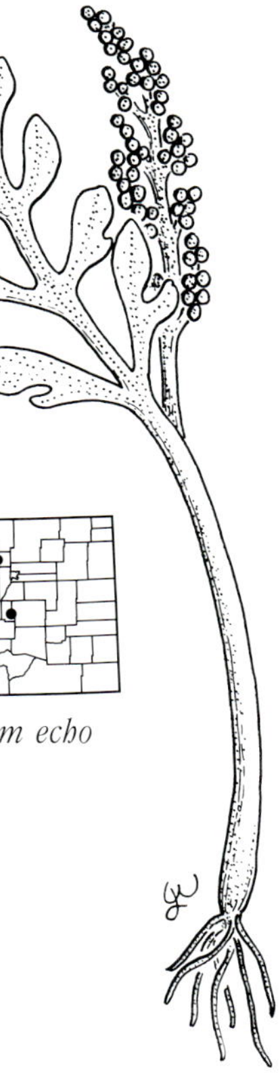

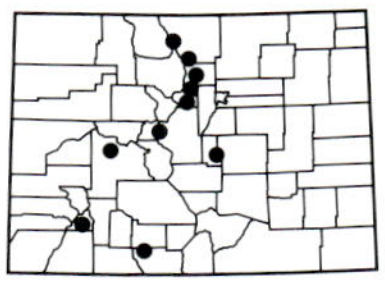

Botrychium echo

MOONWORTS (Adder's-tongue Family)

Botrychium echo W.H. Wagner, ***Botrychium lineare*** W.H. Wagner
Botrychium lunaria (Linnaeus) Swartz, ***Botrychium pallidum*** W.H. Wagner

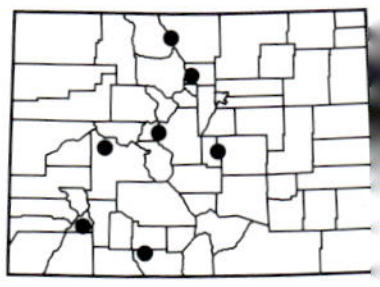

Botrychium pallidum

The small ferns called moonworts are sparsely distributed throughout the mountainous portions of Colorado, but because they are small and inconspicuous they seldom are seen. Most are less than 3 inches tall, although rare giant plants may be as tall as 8 inches. They have underground stems, so all that shows above the ground is a single leaf consisting of a vegetative portion called a trophophore and a spore-bearing sporophore. The sporophore's clusters of spore cases resemble tiny bunches of grapes, so moonworts sometimes are called grapeferns.

Brune

Unlike many rare plants, moonworts often are found in places that have been disturbed in the past. Roadsides and old townsites frequently have large mixed populations of several species. Plants even grow through pavement cracks of some mountain roads. Identification of moonworts is difficult because of subtle differences in the shape of the trophophore segments. *Botrychium lunaria* has distinctive, overlapping, half moon-shaped segments. *Botrychium echo* usually has a thumb-like lobe on the lower edge of its segments, but this can be obscured in dissected leaves, which are common in some populations. *Botrychium lineare* has narrow, straight-sided segments spaced like the cross arms of an old-fashioned telegraph pole. *Botrychium pallidum* is a small species with a distinctive pale green color.

As is currently understood, *Botrychium lineare* is one of the rarest plants in North America. The largest known population is located on Pikes Peak. *Botrychium pallidum* is known from about a dozen Colorado locations. The other species are more widespread, but for most people, finding one is a rare event.

Jennings

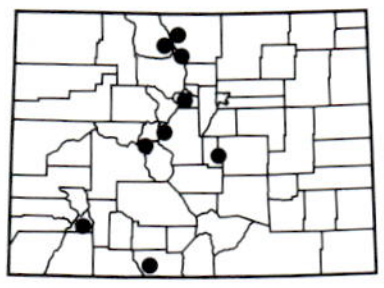

Botrychium lunaria

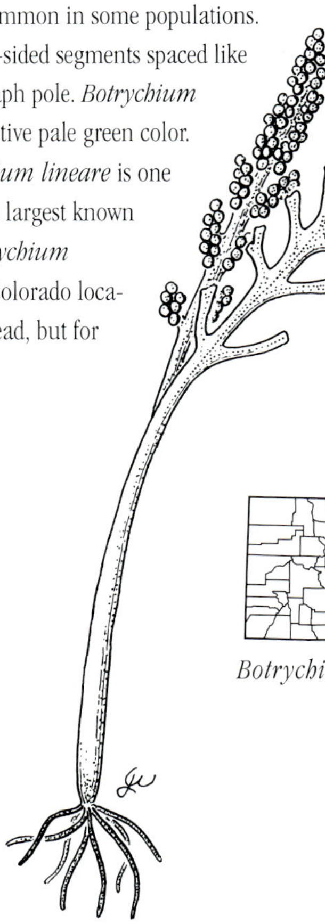

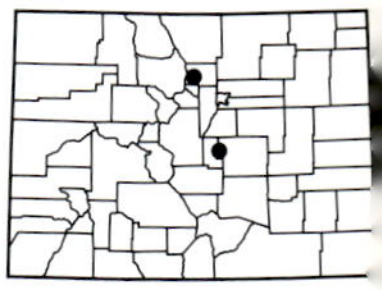

Botrychium lineare

Pictureleaf Wintergreen, Painted Shinleaf

(Wintergreen Family)

Pyrola picta J.E. Smith in Rees

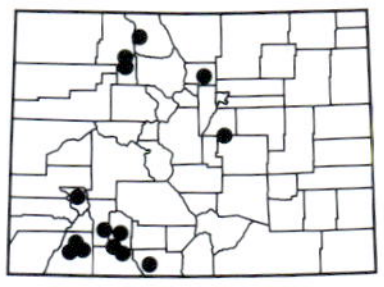

In Colorado, there are several species of wintergreens, but pictureleaf wintergreen is the only one that has ovate leaves with prominent white veins. Others have round or kidney-shaped, uniformly green leaves. *Pyrola picta* has cranberry red flowering stems that are about 5 inches tall. The flowers face downward and are creamy white with a tinge of brownish red. Finding one in bloom is a special treat.

Pictureleaf wintergreen inhabits cool, moist woods, frequently occurring on north- or northeast-facing slopes at elevations from 6,000 to 10,000 feet. Unlike many rare plants, *Pyrola picta* does not seem to show preference for any particular geologic formation. It grows in soil derived from the volcanic rocks of the San Juan Mountains and the granites of the Front and Park ranges.

Pictureleaf wintergreen is widespread in the Sierra Nevada of California, even common in Yosemite National Park. The plant becomes increasingly rare east of the Sierra-Cascade crest, with its easternmost locality in the Black Hills of South Dakota. The reason for its rarity in the Rockies is unknown as there appears to be ample habitat similar to that in California.

Martin

Rattlesnake Fern (Adder's-tongue Family)

Botrypus virginianum (Linnaeus) Holub

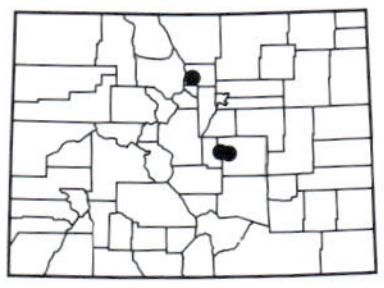

Rattlesnake fern looks more like a typical fern than a moonwort, but it is closely related to moonworts and is sometimes placed in the same genus, *Botrychium*. Its leaf has a feathery, much-divided, triangular vegetative portion and a tall sporophore that bears spore cases resembling small grape bunches. Compared with the moonworts, rattlesnake fern is a huge plant. It can grow as tall as 3 feet, but Colorado specimens are much smaller.

Botrypus virginianum is found in moist forests in nearly every state as well as Mexico, Central America, and Eurasia. In Colorado, it is known from a few wooded canyons along the Front Range in Boulder and El Paso counties. The nearest other populations are in northern Wyoming and the Black Hills of South Dakota.

Martin

Steinkamp

Yellow Lady's-slipper

(Orchid Family)

Cypripedium parviflorum Salisbury
var. *pubescens* (Willdenow) Knight

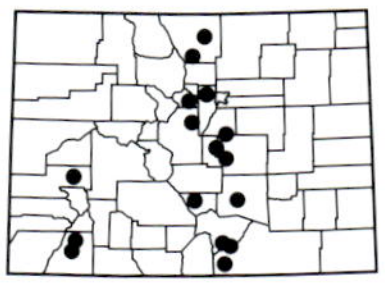

Many people think of orchids as inhabitants of steamy tropical jungles, but native orchids do occur in less spectacular forms in temperate climates. Colorado has fewer than thirty species. Most have small green or white flowers that arouse the interest only of orchid specialists. When examined thoroughly with a hand lens, however, these tiny orchids have all the beauty of their more showy relatives.

Yellow lady's-slipper is the largest flowered orchid in Colorado, and one of the rarest. Sizable populations are known in Larimer and El Paso counties, but elsewhere in the state it is limited to a few scattered patches. Plants can be up to a foot tall but typically are a little shorter. Flowers are large and bright yellow. The lip, shaped like a slipper or moccasin, is about the size of the final joint of the thumb. Plants often are clumped, and seeing several together in full bloom is breathtaking.

Plants similar to the North American yellow lady's-slipper are known from across Europe and Asia. For years, it was debated whether these comprised one variable species or two or more species. Based on recent studies, the Old World and New World plants have been divided into two species, in line with concepts proposed nearly 200 years ago. Colorado plants belong to variety *pubescens*, so named because they are covered with short hairs. Other varieties are found in the eastern United States and Canada.

In Colorado, yellow lady's-slipper occurs at moderate elevations, mostly on the east slope of the Front Range. Increasing urbanization in the foothills and mountains west of Colorado Springs, Denver, and Fort Collins has dramatically reduced yellow lady's-slipper habitat.

Lesser Rattlesnake-plantain (Orchid Family)

Goodyera repens (Linnaeus) R. Brown

Purple Lady's-slipper (Orchid Family)

Cypripedium fasciculatum Kellogg ex S. Watson

Purple lady's-slipper, *Cypripedium fasciculatum*, is the least showy species in a genus noted for showy flowers. Its brownish purple flowers are shaped like miniature moccasins and droop noticeably, making them difficult to see. Usually, there are two to four flowers crowded at the end of a stem. The stem is so weak on some plants that the weight of the flower cluster makes the whole plant bend over, causing the flowers to be sideways or even upside down.

There are two rather large, roundish, glossy green leaves about 2 to 3 inches across that persist until first frost. Plants often are clumped. If snow melts from the subalpine forest habitat on schedule, purple lady's-slipper usually begins to bloom about the third week of June and is finished by mid-July. The plant can be found in lodgepole or spruce-fir forests between about 8,000 and 10,500 feet elevation in northern Colorado. Sites usually are devoid of competing vegetation, with only the orchid growing in pine needle duff or on otherwise bare ground under the trees.

In similar habitat, but farther south in Colorado, one can find the lesser rattlesnake-plantain, *Goodyera repens*. It's not a plantain, but an orchid. The common name is derived from the superstition that the mottled leaves, which have a fancied resemblance to rattlesnake skin, were an effective antidote for snakebite.

Plants in the genus *Goodyera* have a horizontally creeping rhizome just beneath the layer of decaying pine needles and twigs in which they grow. On the surface, the plant displays a rosette of evergreen leaves that usually are mottled with white or light green on dark green. In August, a short stalk bearing several small, white, hairy flowers arises from the center of the rosette. After flowering, the rosette dies. New rosettes are produced elsewhere on the rhizome the following year. As a result, compact colonies a foot or two in diameter can be produced.

Jennings

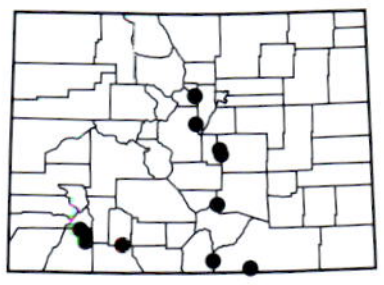

Goodyera repens

Lesser rattlesnake-plantain is a circumboreal species known from across Europe, Asia, Canada, and southward to the high Appalachians and Rockies. In the western United States there are a few populations in Montana and the Black Hills of South Dakota, but the plant skips over Wyoming and reappears in the southern mountains of Colorado, as well as in New Mexico and Arizona. The reason for this unusual pattern of distribution is unknown.

Steinkamp

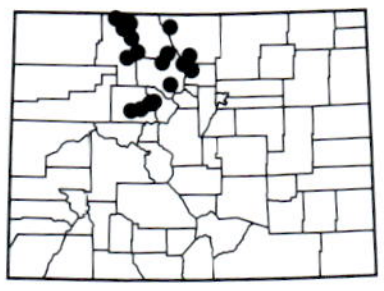

Cypripedium fasciculatum

Steinkamp

Steinkamp

Southern Rocky Mountain Cinquefoil, Giant Cinquefoil (Rose Family)

Potentilla ambigens Greene

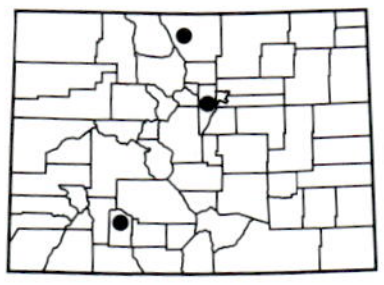

Edward L. Greene was a clergyman-turned-botanist who worked in much of the West during the last three decades of the nineteenth century. In 1899, he found a cinquefoil along Bear Creek near Morrison. A few years later, he described the plant as a new species, *Potentilla ambigens*. It seldom has been seen since that time, and it is known today from only a few scattered populations. In Colorado, it is easiest to find in the vicinity of the old mining town of Creede. Giant cinquefoil also has been reported from northern New Mexico and southern Wyoming.

This species is Colorado's largest cinquefoil, sometimes reaching nearly 20 inches in height. Several stems emerge from a stout rootstock. Flowers are yellow and are typical of cinquefoils. Leaves are more than 8 inches long and usually consist of six or seven pairs of leaflets. The entire plant is covered with long, soft hairs.

Southern Rocky Mountain cinquefoil is especially striking after a frost in the fall, when the leaves turn bright red and make the plant immediately obvious. Some Colorado rare plant species may be considered rare only because they are poorly collected, but the rarity of this large and obvious species has gone unchallenged. The reason for its sporadic occurrence remains unknown.

Selkirk Violet, Great Spurred Violet (Violet Family)

Viola selkirkii Pursh ex Goldie

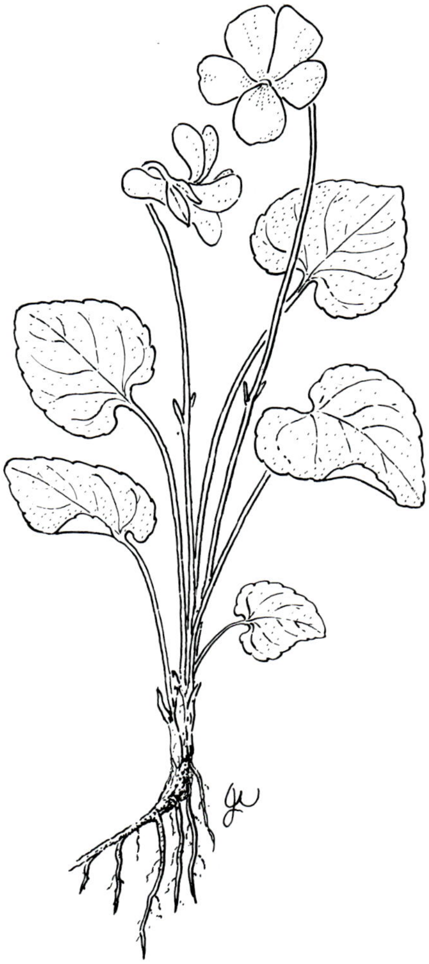

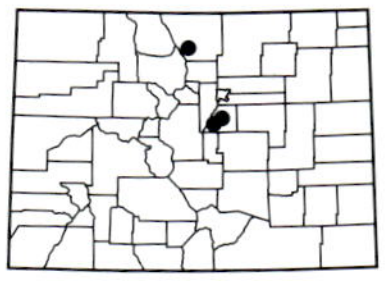

Like many of our rare forest species, this violet is common in Canada but becomes increasingly rare as one travels southward in the Rockies, where it occurs only in environments that meet its specific needs. The nearest populations to Colorado are in the Black Hills of South Dakota. In Colorado, *Viola selkirkii* is known from only two areas: Rocky Mountain National Park, where it was last seen in 1965, and at the east base of Devil's Head in Douglas County, where it was last seen in 1923.

Ruth Ashton Nelson, who spent much of her life studying the plants of Rocky Mountain National Park, was the last person known to have seen this violet in Colorado. Filed with her specimen is a letter describing in considerable detail, practically to the pace, exactly where she found the plants. She described the habitat as very wet woods with plants located on the "insteps" of the alders. Soil gathers around the base of the alders and forms little hummocks that are not quite as wet as the soil on the forest floor. Presumably, this is what she meant by insteps.

In recent years, several botanists have searched for the violet without success. The woods seem to be considerably drier than those described by Nelson in the 1960s; many of the alders and aspens are now dead. The current status of the plant in Colorado is unknown. It simply may have died out.

White Addersmouth (Orchid Family)

Malaxis brachypoda (Gray) Fernald

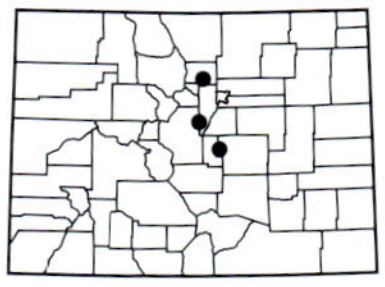

Jennings

All of the Colorado orchids, except the white and green bog orchids and spotted coralroot, are rare and difficult to find. White addersmouth, *Malaxis brachypoda*, takes the prize, however, as the rarest orchid in the state. First reported for Colorado from El Paso County in 1895 and then collected in Boulder County in 1906, the orchid was not found again for 64 years, when the Boulder County population was rediscovered in 1970. A third population was found in Jefferson County in 1971. The El Paso County site never has been rediscovered and may have been destroyed by development. A banner year for white addersmouth in Colorado was 1989 when thirteen plants were found at the two known sites.

Malaxis brachypoda can be found across the Northeast and Canada, where it has much the same distribution as *Listera convallarioides*. In the western United States, however, addersmouth is known only from Colorado and California. The California populations were discovered in the San Bernardino Mountains in 1922, but have rarely been seen since. This is another orchid that is uncommon, yet is afforded no federal protection.

White addersmouth bears a single leaf, the base of which loosely sheaths a very slender, fragile spike of minute, greenish flowers. In the main part of its range, the plant can be up to a foot tall, but it rarely exceeds 3 or 4 inches in Colorado. The flowers are barely larger than a typewritten asterisk (*). Growing along small streams in the lower mountains, white addersmouth usually is found rooted in mosses kept perpetually wet by stream spray. Its small size and totally green coloration make it virtually indistinguishable from the rest of the streamside vegetation.

Maley

Prairie Goldenrod, White Goldenrod-Aster

(Daisy Family)

Unamia alba (Nuttall) Rydberg

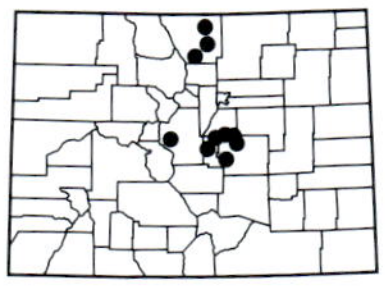

This unassuming, white-flowered perennial is one of those species that give botanists headaches as they try to classify plants. It originally was named *Inula* (*Chrysopsis*) *alba*, but through botanical research it was found to be more closely aligned with the goldenrods (*Solidago*). It didn't quite fit there, either. For one thing, most goldenrods are yellow. So, E.L. Greene established a new genus, *Unamia*, to accommodate it. Because this plant that may have arisen as a hybrid of two other species does not fit well into any other genus, it seems best to have it stand by itself in its own genus. To add to the mystery of the plant, Greene did not explain the derivation of the new generic name. It is speculated that Greene saw the plants as belonging to an unnamed genus and contracted and Latinized the word unnamed to *Unamia*.

Prairie goldenrod is widespread in midwestern woodlands and on the northern prairies. Thousands of years ago, plants now limited to the East and the Midwest extended all the way to Colorado. The area now occupied by the Great Plains was moister, cooler, and forested, at least along streams and rivers. As the glaciers receded and the climate became warmer and drier, the forests retreated and were replaced by extensive grasslands. Some of the eastern flora was able to persist in suitable habitat, particularly at the base of the mountains or in the lower foothills. In Colorado, *Unamia alba* can be found from the Black Forest to Florissant and in a few other locations.

Martin

White-flowered Azalea, Rocky Mountain Rhododendron (Heath Family)

Rhododendron albiflorum Hooker

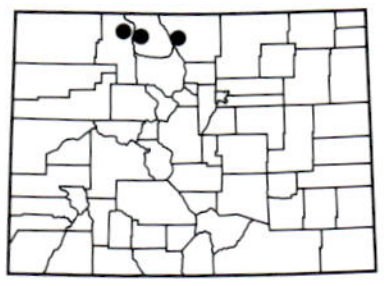

While backpacking in the Mount Zirkel area north of Steamboat Springs in 1911, Edward R. Warren, author of *Mammals of Colorado*, discovered an azalea. Soon after, Aven Nelson of the University of Wyoming described this plant as a new species of *Rhododendron*, naming it *Rhododendron warrenii* after its discoverer. Later comparison of the Colorado plants with azaleas of the far Northwest revealed that *Rhododendron warrenii* was essentially identical to *Rhododendron albiflorum*, a species ranging from Montana to British Columbia and southward into Oregon. Interestingly, the Colorado plants' nearest neighbors are in western Montana some 600 miles away. The azalea is not known in Wyoming.

Since its discovery, a few more populations of *Rhododendron albiflorum* have been found, mostly in the same area but also in the Medicine Bow Mountains of western Larimer County. Requiring cold winters and warm, moist summers, the azalea is a woody plant forming shrubby patches as much as 6 feet high in moist ravines. When in full bloom the plant is spectacular, but this does not happen every year. The sight of a plant bearing dozens of white, creamy, or slightly pinkish flowers is never forgotten and worth the effort it takes to get to an azalea patch.

Native Americans made use of this plant by scraping off the bark, boiling it, and using the decoction as a stomach discomfort remedy. Another use was to reduce the wood to charcoal, which was then crushed and mixed with grease for application to swellings.

Low Northern Sedge, Elegant Sedge (Sedge Family)

Carex concinna R. Brown

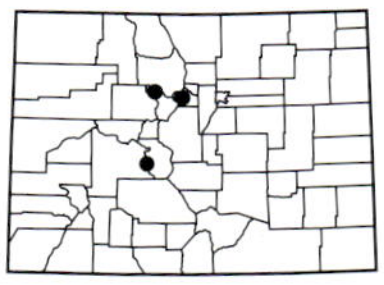

There are many species of sedge considered to be rare or uncommon, perhaps due more to undercollecting of specimens than actual rarity. While many sedges are plants of mountain wetlands, low northern sedge, *Carex concinna*, is a forest species. It is found on cool, moist, mossy forest floors, but not in areas of standing water and saturated soil. Alas, their drab appearance and the lack of showy flowers condemn sedges to be of interest only to sedge specialists, and even knowledgeable botanists may pass over the sedges as "just another grass." Sedges are not grasses, of course. They are in a separate family. Grasses usually have cylindrical or flattened stems which almost always are hollow. Sedges, in contrast, almost always have stems that are triangular and solid. From these characteristics comes a botanical aid to memory: "Sedges have edges . . ."

Sedges seldom are given common names by the general public, so the alternative common name, elegant sedge, is merely a translation of the scientific name. It is debatable whether elegant sedge really fits that description, but beauty is always in the eye of the beholder. It must have been elegant to Robert Brown, who described the species.

Low northern sedge is known from only a few locations in Colorado, typically in subalpine forests at about 10,000 feet elevation. It is somewhat more common in deep forests of the Pacific Northwest. The nearest known populations are in the Wind River Mountains of Wyoming and in the Black Hills of South Dakota.

Jennings

Trillium, Wakerobin (Lily Family)

Trillium ovatum Pursh

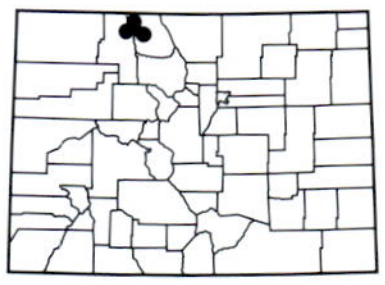

Another plant of the far Northwest, *Trillium ovatum* barely enters Colorado. The main range of the species is from Montana to southwestern Canada and south to California. In the Rockies, it is known from Yellowstone National Park, Wyoming. A few scattered populations straddle the Continental Divide near the state line in Colorado's Park Range and Wyoming's Sierra Madre. Colorado populations usually occur on well-timbered and cool north-facing slopes where the climate resembles that of the Northwest.

Jennings

Just when *Trillium ovatum* was discovered in Colorado has not been determined. The plant was included by Rydberg in his 1906 *Flora of Colorado*, but specimens in local herbaria date only to 1949. It seems likely that the plant was unseen for nearly 50 years.

Trillium is a strikingly beautiful plant. It normally produces a single flower of three large white petals backed by three much smaller green sepals. The flowers gradually turn pinkish with age. Three large, net-veined leaves are arranged in a whorl around the stem, an inch or two below the flower. The genus name is thought to be derived from an old word meaning triplet, probably alluding to this striking foliar arrangement. Taxonomist William A. Weber follows a classification proposed by John Lindley in 1846. The genus *Trillium* is placed in its own family (Trilliaceae).

A number of species of trillium in the eastern United States often are dug from their natural habitat for garden plants. Unfortunately, this also has occurred in Colorado, and a rare plant has become even rarer as easily accessible stands have been removed. Other threats in Colorado include clear-cut timbering and forest fire. Fortunately, the trillium seems to be relatively safe elsewhere in its range.

Spreading Wood Fern (Shieldfern Family)

Dryopteris expansa (Presl) Fraser-Jenkins & Jermy

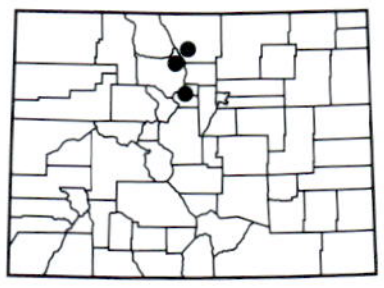

Ferns do not produce flowers, but reproduce by spores instead of seeds. Although without flowers to attract the eye, ferns are familiar to many because of their characteristic leaves or fronds. Spreading wood fern actually is common in northern Europe, southeastern Canada, and in northern Michigan, Wisconsin, and Minnesota. It appears again in the far Northwest, extending southward in the Sierras and the Rockies. In Colorado, at the extreme southern limit of its range, it is to be sought in the familiar cool, moist habitat populated by so many of the other refugees from the Northwest. This fern has been found in a few places above 9,000 feet in Rocky Mountain National Park and once on James Peak. Sites have been described as "rock crevice on cliff," or "observed from across canyon but not visited due to steepness of access," and "no trail, difficult access."

Jennings

Identification of *Dryopteris* ferns can be a very challenging task. As the genus is now understood, in North America there are thirteen fertile species and twenty-nine sterile hybrids. Until it was possible to study chromosomes, there were various confusing interpretations of this group, and spreading wood fern was given different names in different countries. Fortunately, this confusion does not exist in Colorado, where only one other species of *Dryopteris*, the male fern, *Dryopteris filix-mas*, is found.

Grasslands

Grasslands are the predominant vegetation type in eastern Colorado. They cover approximately 40 percent of the state and are part of the larger Great Plains province. On the Western Slope, grasslands are found only as small patches in a woodland-shrubland mosaic. Mountain grasslands occur in high valleys such as South Park or the San Luis Valley. Distribution of natural grasslands is related to topography, climate, fire, grazing, soils, and to combinations of these factors.

Grasslands often occur where soil water availability is too low for forests but sufficient for grasses, some of which can utilize shallow soil moisture from precipitation that falls during the growing season. On the plains, most of the precipitation falls in the spring and summer, just when the grasses experience their maximum growth. In contrast to the neighboring mountains, the soils of plains grasslands mainly originated as sediments in a shallow sea. Rock outcrops, or breaks, and erosion reveal the horizontal nature of the underlying rocks. Many rare plains plants can be found on these breaks, where they are less subject to grazing.

Prior to about 10,000 years ago, forests and woodlands were found at significantly lower elevations, extending far east of the mountain front. As the climate warmed and glaciers receded, strong westerly winds reduced effective moisture, and forests retreated to higher elevations. Most of eastern Colorado then became a shortgrass steppe/mixed-grass prairie mosaic with a few remnants of tallgrass species. This change, however, was not always consistent. Intense droughts and dust bowl conditions occurred, desert vegetation expanded, and sand dunes formed in places. During periods of greater moisture, forests, woodlands, and shrublands expanded back onto the plains.

Even when climate is favorable for grasslands, the species present must respond to the pressures of fire, grazing, and periodic drought. In the absence of fire, mixed-grass and tallgrass prairies can be invaded by woody species. Because of the low availability of fuel in the shortgrass steppe, fire is infrequent and its role in maintaining these prairies is unclear. Plains grasslands generally occupy flat or gently rolling terrain where fires burn with little to stop them. Most native grasses are deep rooted and, thus, fire resistant. They re-sprout almost immediately after a fire. In contrast, woody species generally are restricted to broken terrain, either because of direct fire susceptibility or because their seedlings cannot compete with grasses for moisture and do not become re-established following a fire. Some rare species such as the alpine feverfew, *Bolophyta alpina,* also are found in these rocky plains areas. Well-developed cushion plants such as the rare mountain cat's-eye, *Cryptantha cana*, grow on flat surfaces atop chalky cliffs. In such areas, one has the impression of being on the tundra.

Grazing pressure on the grasslands came primarily from the bison (buffalo), which after about 4,000 years ago roamed the Colorado plains in large numbers. Evolving with the grasslands, these animals grazed preferentially on short upland vegetation, especially on recently burned areas and around prairie dog towns. Prairie dogs often have been characterized as pests that damage grasslands. However, they were a natural part of drier grasslands and played an important role in the fire-grazing cycle. The short, green vegetation within prairie dog colonies usually cannot support fire, contributing to a patchy, burned-and-unburned environment across the prairie.

Most Colorado grasslands are used heavily by humans. The major urban areas of the state are on the plains, with buildings and landscaped vegetation replacing prairie. The native plains vegetation has been altered even more extensively by agriculture. Virtually every flat area with arable soils has been converted to cropland. Most native grassland species are long-lived perennials that reproduce poorly from seed under natural conditions. So, plowed sites may need to lie fallow for more than 50 years to begin to resemble the original, unplowed state. The semiarid environment of eastern Colorado makes restoration or recuperation of grasslands extremely slow.

Lowland grasslands often have been used for hay, with native species frequently being replaced by introduced grasses. If used for grazing, native lowland grasslands are damaged quickly because they did not evolve with heavy grazing pressure. Several of the rarest Colorado plants, including the tulip gentian, *Eustoma grandiflorum,* and Colorado butterfly weed, *Gaura neomexicana* ssp. *coloradensis*, are associated with such wet lowland prairies. Shortgrass and mixed-grass prairies, on the other hand, have moderate to high grazing resilience and may recover more quickly from grazing damage if the practice is not sustained indefinitely.

Only limited areas of native grassland are being preserved. Because human use has been so extensive and intensive, only a very few prairie examples remain. The Pawnee and Comanche national grasslands contain small areas that retain their pre-settlement conditions. The Tallgrass Natural Area, part of Boulder's Open Space program, protects small parcels of tallgrass prairie; some Western Slope grasslands are preserved in national parks or monuments. For the most part, however, Colorado's grasslands have received little attention with respect to preservation or restoration.

GRASSES (Grass Family)

There are a great many grass species in Colorado grasslands. Some are widespread, such as blue grama or buffalo grass. Others such as big bluestem have restricted distribution in Colorado. A few are truly rare.

Prairie fescue, *Festuca campestris,* was first noted in eastern Washington. It is known in Colorado from a single occurrence near the Spanish Peaks along the Huerfano-Las Animas county line. It is similar to Hall fescue but does not have rhizomes. It also is similar to rough fescue, *Festuca scabrella,* and some botanists consider Hall fescue and prairie fescue to be variants of the widespread rough fescue. Rough fescue is not known from Colorado. These and several other species are similar enough that worldwide study of the complex will be necessary before relationships can be resolved.

Forktip three-awn, *Aristida basiramea,* is fairly widespread on the eastern Great Plains but rare west of the 100th Meridian (about Dodge City, Kansas). This annual grass blooms in late summer. The lemma, the outer bract of the grass floret, has three stiff appendages called awns. The feature that distinguishes forktip three-awn from its relatives is its central awn, which is longer and spirally coiled at the base. In the eastern Great Plains, this is a species of disturbed areas. In Colorado, it is known from a gravelly site on Rocky Flats south of Boulder. Otherwise, sites in Colorado are in sandy soils on sandstone outcrops.

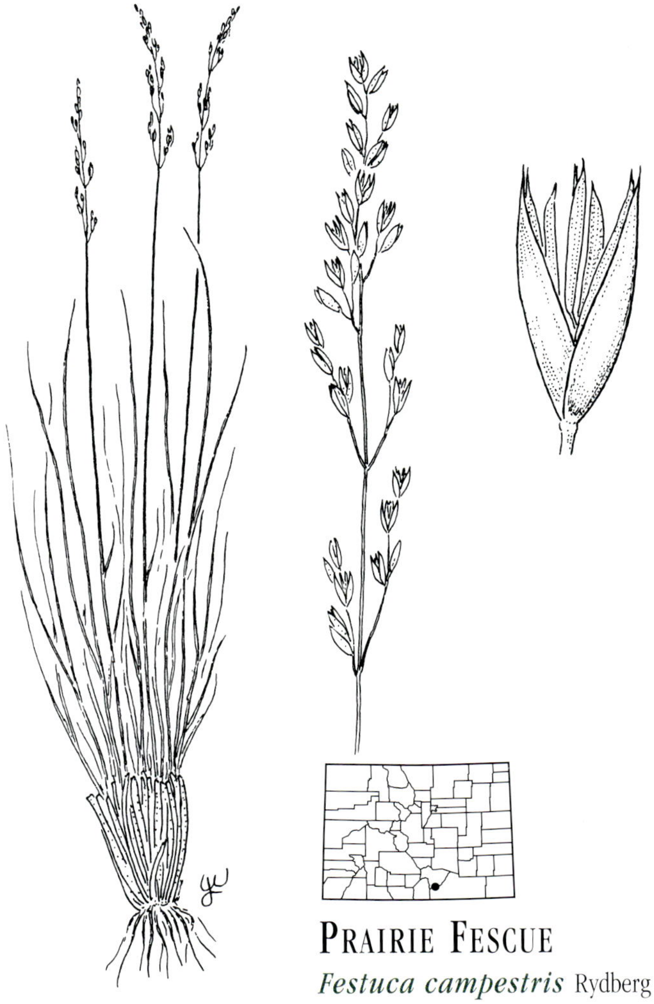

PRAIRIE FESCUE
Festuca campestris Rydberg

Crawford

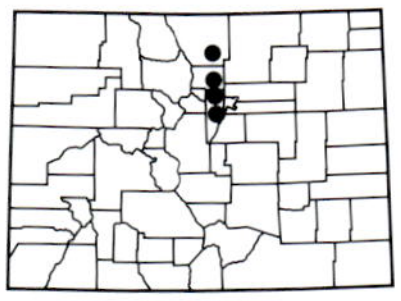

FORKTIP THREE-AWN
Aristida basiramea Engelmann

In 1862, pioneer botanists Elihu Hall and J.P. Harbour collected Hall fescue, *Festuca hallii,* near the north end of South Park. However, the population they sampled never has been relocated. In 1956, William A. Weber of the University of Colorado was shown a patch of fescue in tundra high on the north slope of Cameron Mountain in northern Larimer County. It was Hall fescue, rediscovered after 94 years. The grass forms a sod due to the presence of rhizomes, underground stems that root at the nodes, eventually producing a large clump of plants. At the Cameron Mountain site, Hall fescue is co-dominant with *Kobresia myosuroides,* a member of the sedge family that usually indicates mature soils on relatively dry but peaty tundra.

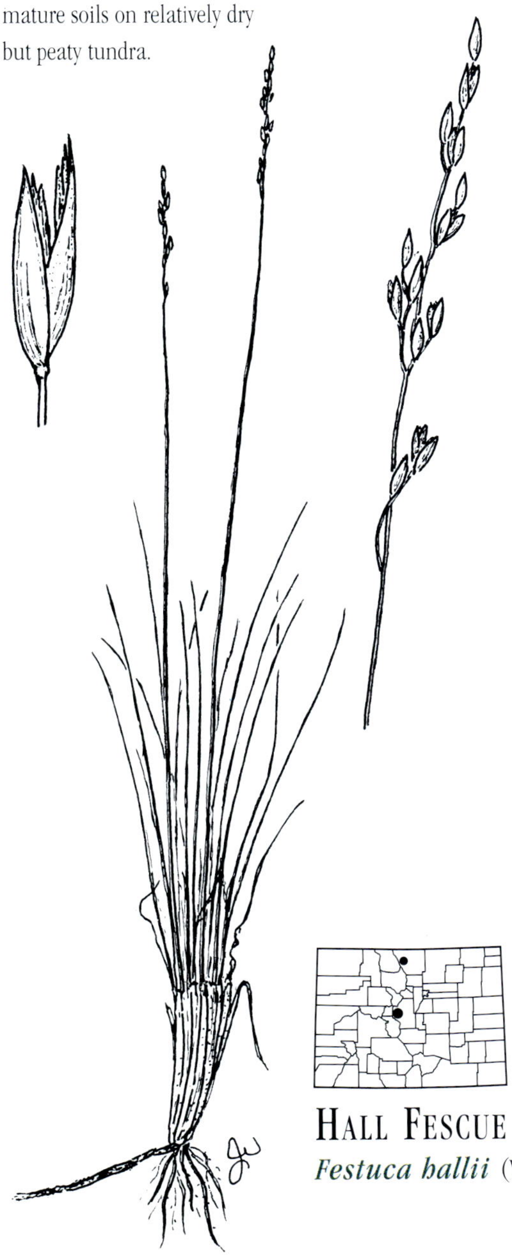

Hall Fescue

Festuca hallii (Vasey) Piper

Jennings

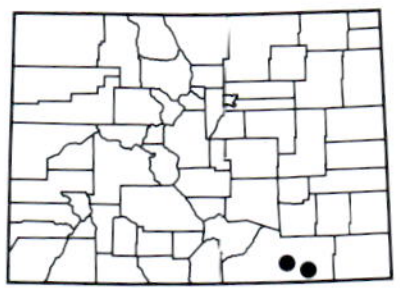

Springfield Silver Beardgrass

Bothriochloa springfieldii (Gould) Parodi

In 1993, during thesis research on the Mesa de Maya in Las Animas County, Dina Clark discovered an unfamiliar grass. Comparison with herbarium specimens showed the grass to be Springfield silver beardgrass, *Bothriochloa springfieldii,* previously unknown in Colorado. In September of 1995, during a Colorado Native Plant Society field trip in Purgatoire Canyon, Clark was surprised to see the grass a second time. These two locations are the northernmost sites for the species, which otherwise is known from scattered sites in western Texas and eastern New Mexico. This is a most attractive grass, with long silky hairs on the flowering parts of the plant and spreading hairs at the nodes on the stem, much like a miniature ruff. The plant is 2 to 3 feet tall and matures in late summer. It is similar to the common and widespread *Bothriochloa laguroides,* and grows with that species in Purgatoire Canyon. When seen together, however, the differences are apparent.

Jennings

Martin

DWARF LEADPLANT (Pea Family)

Amorpha nana Nuttall

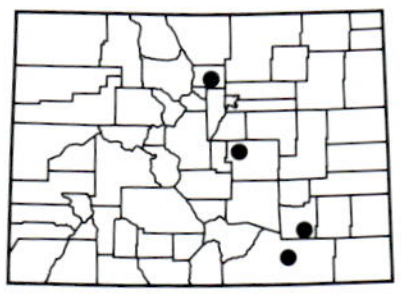

Dwarf leadplant, *Amorpha nana*, is a low shrub, usually about 2 feet tall. In Colorado, it is found on dry prairie at the base of the mountains, most frequently on relatively cooler north-facing slopes. It is one of the few shrubs that occur in the grasslands of central North America. Plants are thinly scattered from Boulder County to the Black Forest region of El Paso County, with a few plants known from Las Animas and Otero counties east of Walsenburg and Trinidad.

The plant was collected and described by Thomas Nuttall, a professor at Harvard University who ventured up the Missouri River in the early nineteenth century. He probably found the plant in North or South Dakota, where it is relatively common. As one travels south on the Great Plains it becomes increasingly scarce in Nebraska, Kansas, and Colorado.

Dwarf leadplant is easily recognized by its aromatic foliage and catkin-like cluster of small purple flowers. The leaves contain numerous resin pockets that appear as small, translucent dots when the leaf is viewed against the light. The name of the genus is derived from the Greek word *amorphos,* variously translated as shapeless or deformed. This is an allusion to the fact that four of the five petals typical of the pea flower are missing.

YELLOW STARGRASS (Yellow Stargrass Family)

Hypoxis hirsuta (Linnaeus) Coville

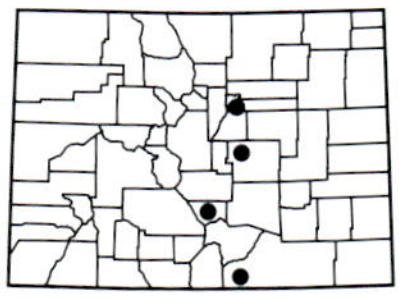

Diligent search of moist portions of the Colorado prairies, particularly at the base of the mountains, might disclose a few plants of yellow stargrass, but it is unlikely. Although this plant is quite common in the Midwest, East, and South, it has been reported from Colorado only five times in 100 years, never twice from the same place. The nearest known localities to Colorado are in central Nebraska. Yellow stargrass, *Hypoxis hirsuta*, has not been reported from Colorado since 1959. It must be considered extirpated in Colorado unless we learn otherwise.

From a nearly spherical corm (the enlarged, solid, bulb-like base of the stem) about an inch in diameter arises a rosette of narrow, linear leaves. These leaves can be as much as 2 feet long, although they are much smaller on the few Colorado specimens. The leafless flowering stems arise from the center of the rosette and have two to seven yellow flowers with three petals and three sepals, all alike. These characteristics are typical of the lily family, with which yellow stargrass sometimes is grouped. The minuscule seeds are covered with tiny spines and look like a miniature hedgehog when viewed with a microscope.

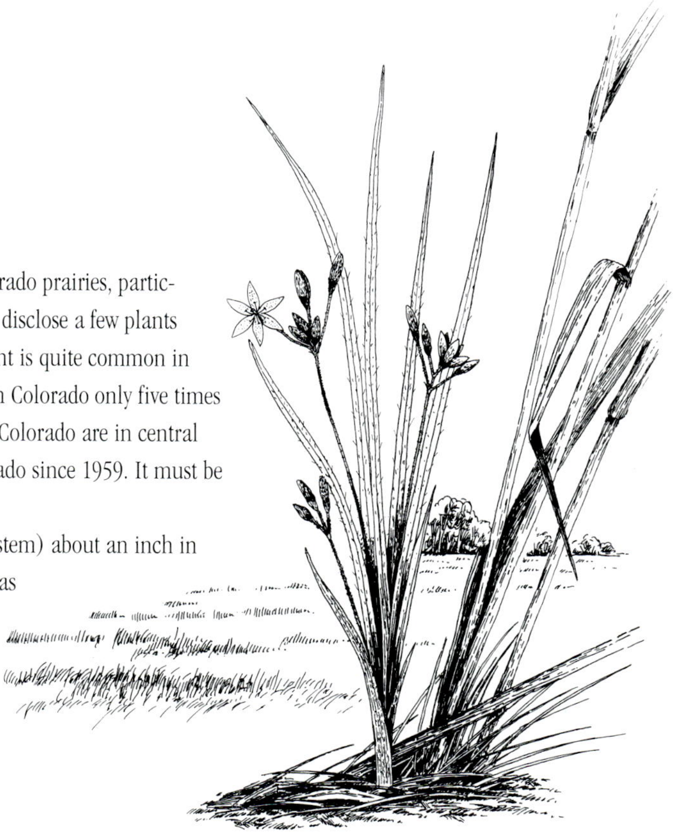

SANDHILL GOOSEFOOT (Goosefoot Family)

Chenopodium cycloides A. Nelson

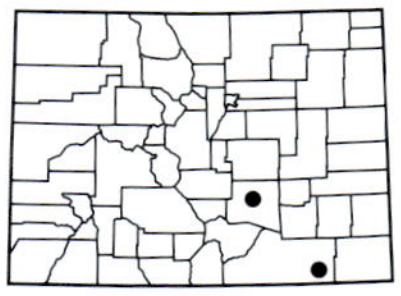

Many times, we cannot say with certainty that a plant is truly rare. Botanical science is dependent upon study of specimens collected over many years and stored in herbaria, usually at major universities. The West is vast, herbaria are few, and collecting is spotty, particularly in areas far from the botanist's home herbarium. Also, botanists frequently do not venture into areas perceived to be "uninteresting," or they do not collect "weeds."

The limited collections of *Chenopodium cycloides* may be a result of such problems. Its main range covers the area where Colorado, Kansas, Oklahoma, New Mexico, and Texas all come together in a jumble of state lines. The area is distant from any major university. Sandhill goosefoot inhabits sandy areas on the plains, which are hot, dry, and dusty during the summer blooming season. It is an annual and only appears when conditions are just right. As a result, it is a serendipitous event when a botanist ventures into such an area, finds the goosefoot, and makes a collection.

Our present knowledge is that in good years, sandhill goosefoot is abundant in the sand hills of the Cimarron National Grasslands in extreme southwestern Kansas and 300 miles to the south in the Monahans sand hills west of Odessa, Texas. However, populations of this species appear to be small and localized elsewhere. It has been collected a few times in New Mexico and four times in Colorado.

Sandhill goosefoot can be up to 3 feet tall. It bears a large number of inconspicuous flowers less than ⅛ inch across. Its most distinctive feature is that each flower has a small, wavy-margined collar that enlarges and spreads as the fruit develops.

Crawford

Locklear

COLORADO GREEN GENTIAN (Gentian Family)

Frasera coloradensis (C.M. Rogers) D.M. Post

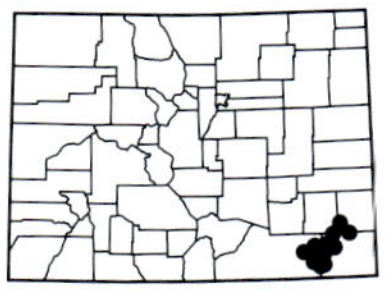

In 1947, while doing field collecting for his doctoral thesis, Claude M. Rogers discovered the Colorado green gentian. Known only from a few populations in Baca, Las Animas, Bent, and Prowers counties in southeastern Colorado, the gentian is found with typical grassland plants and occasionally with juniper (*Juniperus monosperma*).

Frasera coloradensis produces several branched, flowering stems up to a foot tall. The plant grows for several years as a vegetative rosette before flowering and fruiting, and apparently dies afterward. Leaves and floral bracts have distinct white margins. The four-petaled flowers are whitish or greenish with purple dots, measure about ¾ inch across, and appear rather ordinary for a family known for showy flowers.

The greatest threat to Colorado green gentian appears to be from cattle grazing in mid- to late summer. If little other forage is available, the gentians may be consumed before they can produce mature seed.

Weber

COLORADO BUTTERFLY WEED (Evening-primrose Family)

Gaura neomexicana **Wooton ssp.** ***coloradensis*** **(Rydberg) Raven & Gregory**

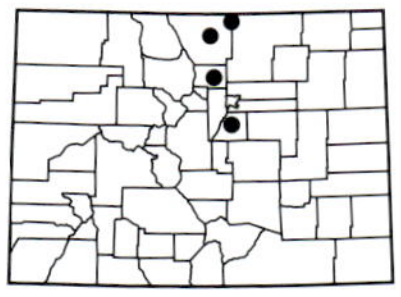

In moist prairie meadows along the mountain front from Castle Rock, Colorado, to Cheyenne, Wyoming, in the transition zone between wet streambottom and rich floodplain areas, a few scattered populations of Colorado butterfly weed have been found. It was first reported in 1895 by a Colorado State University botanist in what is now the city of Fort Collins. Much habitat where it would be likely to occur has been altered for agricultural use, usually hay production, or destroyed with increasing urbanization along the Front Range. A few small populations were recorded through the years elsewhere in northern Colorado. Finally, in 1994, the plant was rediscovered in northern Larimer County. Meanwhile, intensive searching by Wyoming botanists uncovered several populations of *Gaura neomexicana* ssp. *coloradensis* just across the state lines in southeastern Wyoming and western Nebraska.

This large, obvious, but rather weedy-looking plant grows to 4 feet tall, branching numerous times from the main stem to create a pleasing candelabra effect. The white flowers, which are about ¾ inch across and four petaled, quickly turn pink or rose as they fade. Like those of most other members of the evening-primrose family, blossoms of Colorado butterfly weed open in the evening. By noon the next day they have faded, each flower lasting only one night. The blooming period, however, extends from July into September as flowering proceeds from the lower end to the top of the long stalks.

PRAIRIE MOONWORT (Adder's-tongue Family)

Botrychium campestre **Wagner & Farrar**

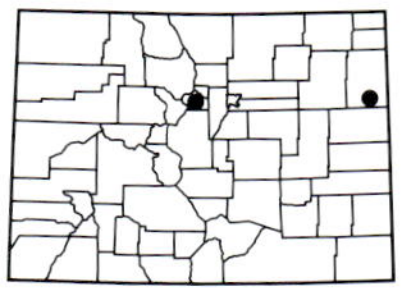

Moonworts are not often found in grasslands, but over the years a few were found and usually identified as *Botrychium minganense*. These prairie plants recently have been recognized as a distinct species, *Botrychium campestre*, the prairie moonwort. Prairie moonwort is found over a wide area of the Great Plains and on dunes by the Great Lakes. In Colorado, a small population occurs in the loess prairie natural area near Bonny Reservoir in Yuma County. These plants emerge from the ground in May before the prairie grasses start their growth. They probably disappear in June.

One peculiarity of this species is its production of minute underground gemmae, which can detach from the parent and mature into new plants. It has been suggested that the presence of gemmae in other species indicates a relationship to *Botrychium campestre*. *Botrychium echo*, a subalpine species, has gemmae and is thought to have originated as a hybrid between *Botrychium campestre* and *Botrychium lanceolatum*. Where did this hybridization occur? This question led Colorado naturalist Peter Root to search for the prairie moonwort in mountain habitats. He has found one plant at Echo Lake in Clear Creek County at 10,000 feet elevation. Continued searches in mountain and grassland habitats may reveal other populations.

Prairie Parsley (Parsley Family)

Musineon tenuifolium Nuttall ex Torrey & Gray

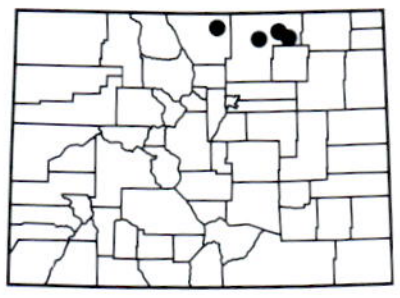

Prairie parsley, *Musineon tenuifolium*, is limited to the region on the high plains where Colorado, Nebraska, and Wyoming join. In Colorado, it is known from dry prairie on the Pawnee National Grassland.

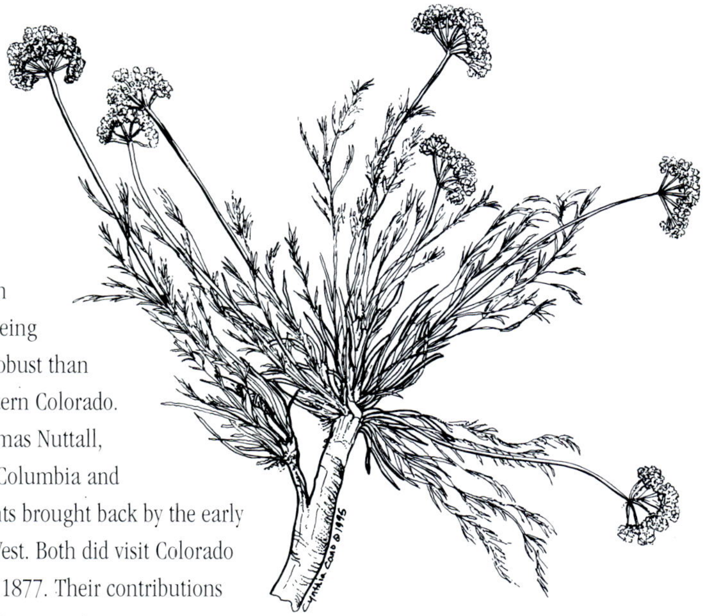

During May and June, the plants produce white or slightly yellowish flowers, typical of many of the parsleys. Leaves are all basal rather than being borne on a visible stem. Plants can be up to a foot tall, somewhat more robust than many of the plants adapted to the harsh prairie environment of northeastern Colorado.

This species was made known to science through the efforts of Thomas Nuttall, who collected plants along the Oregon Trail in the 1830s. John Torrey of Columbia and Asa Gray of Harvard collaborated in formally describing many of the plants brought back by the early botanical explorers, although they personally collected very little in the West. Both did visit Colorado in 1872 (Gray in August and Torrey in September), and Gray returned in 1877. Their contributions to botany are commemorated by Grays and Torreys peaks, two 14,000-foot mountains near Loveland Pass.

M. Denham

M. Denham

Plains Ragweed, Streaked Ragweed (Daisy Family)

Ambrosia linearis (Rydberg) Payne

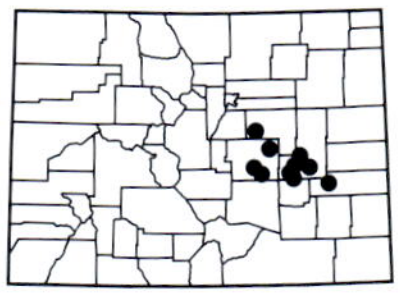

Plains ragweed, *Ambrosia linearis*, is a rarely collected endemic of the plains of Colorado. First found by DeAlton Saunders in 1893 near Calhan, it went unreported for 40 years before it was collected independently by three botanists in Elbert, El Paso, and Lincoln counties. In the 1970s and '80s, it has been found by botanists returning to those areas or searching for new county records. Originally, plains ragweed grew at the margins of playa lakes. When engineers with a penchant for straight lines cut road and railroad rights-of-way through the playas, plains ragweed willingly moved on to roadside ditches and railroad embankments. It is there that plains ragweed most often is encountered today, in sandy or gravelly, seasonally moist soils.

W.W. Payne, an expert on the genus *Ambrosia,* wrote in 1964 that plains ragweed was "a very distinctive species, inexplicably rarely collected." The erect, shrubby plant arises from a somewhat woody base and stands about a foot tall. Leaves are bright green on the upper surface, but appear white on the lower surface due to the presence of numerous minute hairs. Each *Ambrosia linearis* plant bears separate male and female flowers. The male flowers are yellow and a little showy; the female flowers are inconspicuous.

Ambrosia is fabled to be the food of the gods. When eaten mixed with nectar it is reputed to make one immortal. As far as we know, no Colorado botanist has verified this.

Steinkamp

TULIP GENTIAN, SHOWY PRAIRIE GENTIAN (Gentian Family)

Eustoma grandiflorum (Rafinesque) Shinners

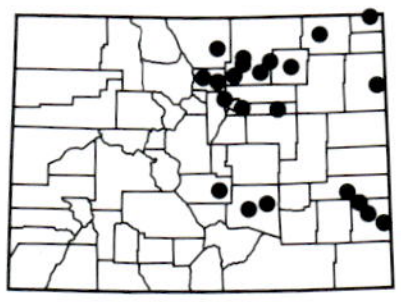

Formerly found on the western half of the Great Plains from South Dakota south through Texas into Mexico, populations of tulip gentian have been reduced greatly since settlement. Its habitat has been reduced to the point that tulip gentian, *Eustoma grandiflorum*, now is rare over much of its former range. The story is the same in Colorado. Originally known from twelve Colorado counties and seen in a vacant lot in Denver as recently as 1981, the nearest tulip gentian populations now are at least 35 miles from Broadway and Colfax, the center of the city.

Tulip gentian requires a fairly high water table in moist open fields and meadows underlain by sandy alluvial soils. Unfortunately, on the dry plains of Colorado this is prime grazing or hay production area. Not only agricultural use but gravel mining and urbanization have eradicated most populations near the eastern mountain front.

Eustoma grandiflorum is one of the showiest rare plants in Colorado. Seeing it for the first (or even tenth) time always evokes delight. Standing about 1 to 2 feet tall with pale gray-green leaves, the plant produces spectacular clusters of two to six lavender, blue-purple, or pink flowers about 2 inches across. Plants tend to grow in profuse groups, adding more and more flowers to the display.

This species is being bred intensively in Japan and the United States for horticultural purposes. It currently is grown in greenhouses around the world for both potted plant culture and cut flower production. Whether it will continue to exist in the wild in Colorado is unclear.

RICHARDSON ALUMROOT (Saxifrage Family)

Heuchera richardsonii R. Brown

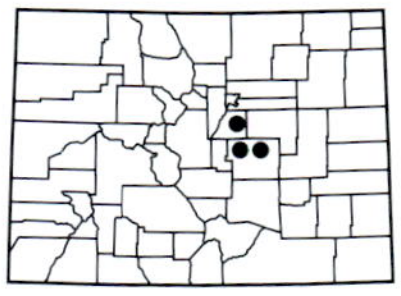

Robert Brown, who described this species, was an early nineteenth-century librarian, botanist, and the first "keeper of botany" at the British Museum. He named this alumroot for his contemporary Sir John Richardson, who was botanist and zoologist attached to Sir John Franklin's expedition to the American Arctic. Many northern plants were brought to England by Richardson, and his name is commemorated in a number of species.

Although *Heuchera richardsonii* does not reach the Arctic, it is widespread on the extreme northern Great Plains in Alberta, Saskatchewan, and North and South Dakota. Like other rare grassland species, it becomes increasingly scarce on the southerly portions of the plains, where prolonged hot, dry periods often occur. In Colorado, Richardson alumroot largely abandons the plains, taking refuge from the July heat in grassy openings in the Black Forest northeast of Colorado Springs. As the city expands, this area has become increasingly developed with negative consequences for the plant. However, one small population is protected in Castlewood Canyon State Park.

The saxifrages are valued as rock garden plants. Some are showy, with red or yellow flowers. Plants in the genus *Heuchera* are usually rather drab, with greenish or whitish flowers. Richardson alumroot has a hairy flowering stem up to 2 feet tall arising from a cluster of basal leaves. The flowers are greenish, large for an alumroot (about ½ inch across), and somewhat asymmetrical.

Maley

WYOMING FEVERFEW (Daisy Family)

Bolophyta alpina Nuttall

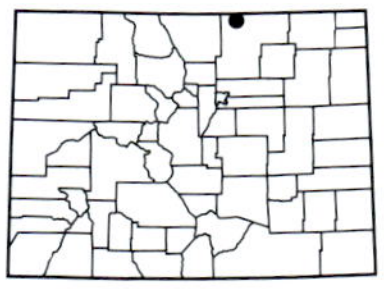

The driest parts of the plains grasslands in northern Colorado and eastern Wyoming, particularly gravelly or barren slopes without much competition from other plants, are home to the Wyoming feverfew. This mat-forming plant shows very little above ground level, but has a large, thick, branched root that stores enough food to keep the plant going through the lean times. Small leaves about ¾ inch long are borne in tufts at ground level. Flowering heads of whitish or greenish flowers appear in late April or early May, well before new leaves appear. Thus, because the flower heads are buried among the previous year's dead leaves they are almost invisible. Later, the appearance of new green leaves changes the plant's appearance completely.

Steinkamp

Wyoming feverfew is another one of the "lost plants." First seen by the pioneer botanist Thomas Nuttall in 1834, it was not reported again for more than 100 years. A number of populations since have been found in southeastern Wyoming and in 1977, William Harmon of the University of Northern Colorado in Greeley found the plant at two remote localities in Weld County, Colorado. In recent years, a few more populations have been found in the same general area.

Even though *Bolophyta alpina* is unprotected on the grasslands, it grows on sites with very little forage and is bypassed by grazing animals.

Locklear

Martin

DWARF MILKWEED (Milkweed Family)

Asclepias uncialis Greene

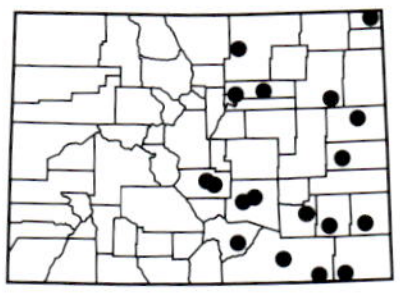

In 1880, Edward L. Greene found a very small milkweed in the vicinity of Silver City, New Mexico. He named the plant *Asclepias uncialis*, the specific name meaning an inch long. The name refers to the size of the plant, a tiny species in a family commonly known for large plants with showy flowers. The entire plant is usually less than 3 inches tall. The lower leaves are ovate and the upper ones are long and narrow. Leaf margins and veins have small, incurved hairs. Flowers are purplish rose. Dwarf milkweed's seeds, like those of other milkweeds, are equipped at one end with a tuft of silky hairs that serves as a parachute and enables the seed to be blown around by the summer wind.

During a recent study of the species, James Locklear of the Nebraska Statewide Arboretum searched herbarium collections of more than twenty botanical institutions. He located specimens collected from Arizona, New Mexico, and Colorado, and found that most were collected in the nineteenth century. Returning to the sites of these and other historical collections, he was able to find plants at some locations, but the largest population was composed of only a few individuals. In 1995, a population with several hundred plants was found in Las Animas County.

Dwarf milkweed plants tend to occur at the base of sandstone mesas or canyon walls, in the barren spaces between clumps of blue grama or buffalo grass where the accumulated debris and soil slope away from the cliff. There are many sites with this description in New Mexico and eastern Colorado, so the reason for so few observations of this species in the past century is unclear. Is it vanishing? Or does its inconspicuous appearance hamper efforts to find it?

As evidence of the latter idea, consider the experience of a 1995 Colorado Native Plant Society field trip group led by Dr. Locklear. After reviewing the description and seeing photographs and herbarium specimens of *Asclepias uncialis*, about twenty persons began searching a hillside where Dr. Locklear had found a few plants in a previous year. It was more than 2 hours later that the first plant was located. After a hands-and-knees view of this plant everyone began searching even more intensively around it. Eventually, about twenty flowering plants were located in an area of several hundred square feet. Despite such additions to our knowledge of dwarf milkweed, the small number and size of its known populations give cause for continued concern.

CHAFFWEED (Primrose Family)

Centunculus minimus Linnaeus

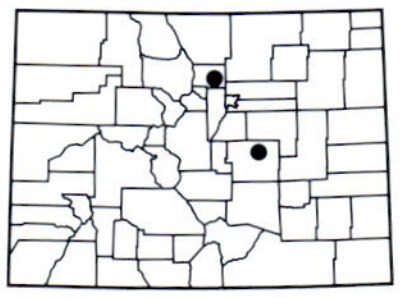

Most primroses are incredibly beautiful plants and many primrose species are highly prized in cultivation. In a never-ending effort to present the story on rare plants, the authors of this book have managed to find the world's least attractive primrose. Sure enough, it's a rare plant with the equally unattractive common name of chaffweed.

Centunculus minimus is extremely inconspicuous, so weak and slender that it requires the surrounding vegetation to hold it erect. It is associated with sedges and rushes on floodplains at the base of the mountains, where it occurs on little tussocks and sometimes forms clumps or mats. If the surrounding vegetation is insufficient to keep it erect, chaffweed leans over and begins to root along the stem. In July, chaffweed produces tiny, pink, solitary flowers in the leaf axils (the upper angle at the point of attachment of the leaf to the stem). The plants themselves are usually less than 4 inches tall. Stems are the diameter of heavy thread and leaves are only about ¼ inch long.

There are no known sites for chaffweed from the driest parts of the Great Plains, but it is present in the cooler or moister portions, such as northeastern Wyoming, northern North Dakota, eastern Nebraska, and eastern Oklahoma. In Colorado, it has been found in wet swales in Boulder and El Paso counties.

MOUNTAIN CAT'S-EYE, MAT CRYPTANTHA (Borage Family)

Cryptantha cana (A. Nelson) Payson

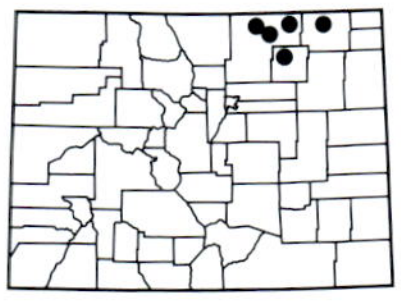

The plains in Colorado are not just a vast unbroken expanse of grasslands. Here and there, geologic and weathering conditions have formed rocky breaks. At Pawnee National Grassland in Weld County there are cliffs high enough to provide nesting sites for raptors. At the top of these cliffs, in a strip just a few feet wide where the soil is too thin or winds are too strong to permit the development of true grasslands, there are some unique cushion plants and dryland species reminiscent of alpine habitats.

Steinkamp

Mountain cat's-eye, *Cryptantha cana,* forms large, low-growing clumps with numerous white-flowered heads standing a few inches above the leaves. Stems are clustered densely and branch at the ground surface above a woody rootstock. Old leaves and leaf bases persist beneath the current year's growth and gradually decompose, forming new soil and frequently contributing to a mounded appearance for the plant. The range of this species is western Nebraska, eastern Wyoming, and northeastern Colorado.

Shrublands

Shrublands are ecosystems dominated by plants with persistent woody stems, but without a large central trunk. Shrublands constitute much of the vegetation on Colorado's Western Slope and in the intermountain parks, but they are less common on the plains. Because shrubs and trees obtain most of their water from deep soil moisture, they can make use of winter soil moisture recharge. Grasses, however, are dependent upon growing season precipitation. Total annual precipitation is similar on both sides of the mountains, but it occurs mostly in the fall and winter on the Western Slope, limiting the development of grasses dependent on summer moisture.

After the end of Pleistocene glaciations about 10,000 years ago, the climate became warmer and drier, allowing southern desert vegetation to move northward. Shrublands replaced grasslands, and grasslands replaced coniferous forests. That trend was reversed by 4,000 years ago, and further fluctuations have occurred to the present. Bison (buffalo) expanded their range to the northern Great Basin about 3,000 years ago. However, their distribution was limited by the availability of permanent water and several other factors, and the Great Basin herds were never as large as those on the Great Plains. Climatic changes reduced their habitats and numbers so that by the early nineteenth century, the bison of the Great Basin were nearly gone.

Most people envision vast expanses of sagebrush when they think of shrublands. However, several types of shrublands can be found on the Western Slope and in intermountain parks. Mat saltbush shrublands dominated by *Atriplex cuneata, A. corrugata,* or *A. gardneri* occur on the driest sites, where vegetation covers less than 10 percent of the surface. Soils of these areas are composed of heavy, shale-derived alkaline clays, which often form areas termed badlands or barrens. The barrens include so many rare plants that they are given a separate chapter in this publication. Greasewood (*Sarcobatus vermiculatus*) shrublands are found in the San Luis Valley and other areas where there are saline-alkaline soils with high water tables. Cold-desert shrublands occur on dry shale or sandstone hillsides where saltbush (*Atriplex confertifolia*) and shadscale (*Atriplex canescens*) are dominant. Even where sagebrush (*Artemisia* spp.) prevails, a number of species and subspecies characterize different sites, depending upon slope, soil depth and texture, height of the water table, and other features. On both sides of the Continental Divide, mixed mountain shrublands are found on sites that are too dry for forest but more moist than pinyon-juniper woodlands. These shrublands, which often include mountain mahogany, Gambel oak, serviceberry, and chokecherry, may form a discontinuous transition between forest and grassland or between forest and sagebrush shrubland. Rare plants can be found in a number of these shrubland types. For instance, Cronquist milkvetch, *Astragalus cronquistii,* and Colorado desert-parsley, *Lomatium concinnum,* are found in the driest saltbush communities. Skiff milkvetch, *Astragalus microcymbus,* is found in sagebrush.

Like many other plant communities, the shrublands have been altered in a variety of ways for use by humans. At the base of the eastern mountains, foothill shrublands have been replaced by houses and landscaping. Mixed mountain shrublands are used for grazing and oak stands have been harvested for firewood and posts. Because soils beneath oak are rich in humus, some of these communities have been cleared for cropland. A few western shrublands have been converted to farming and fruit tree cultivation, especially in the Grand and San Luis valleys, and many of the driest badland sites are favorite off-road vehicle areas. Mostly, though, Western Slope shrublands are threatened by attempts to convert them to grasslands through herbicide spraying, chaining, and fire. Heavy grazing pressure results in shrubland changes that are not immediately apparent. Because cattle make little use of most shrubs, the dominant plants of the community may appear relatively unaffected. However, the grasses and other non-woody components of the shrubland are removed under heavy grazing. These changes may be irreversible under present climatic conditions.

Preservation of shrublands has not been a high priority, primarily because many of these areas have little perceived aesthetic appeal. However, the number of rare and endangered plants found in shrubland communities is high, and these habitats deserve more attention. A few shrubland areas are receiving some protection through their occurrence in national parks and monuments or their designation on other federal lands as Areas of Critical Environmental Concern.

Jennings

Jennings

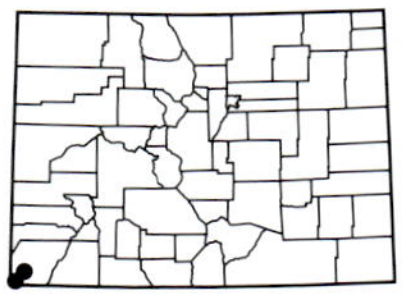

Astragalus cronquistii

CRONQUIST MILKVETCH (Pea Family)
Astragalus cronquistii Barneby
TWISTED MILKVETCH (Pea Family)
Astragalus tortipes Anderson & Porter

Described by Rupert Barneby in 1964, Cronquist milkvetch, *Astragalus cronquistii*, was named in honor of its discoverer, Arthur Cronquist, a noted botanist associated with the New York Botanical Garden. Found first in the Comb Wash area of San Juan County, Utah, this plant's range is limited to that county plus adjacent Montezuma County, Colorado. The few known populations in Colorado are in desert shrub communities dominated by saltbush (*Atriplex* spp.).

Cronquist milkvetch sometimes is found with *Eriogonum clavellatum* and *Astragalus tortipes*, both rare plants in their own right. Twisted milkvetch, *Astragalus tortipes*, occurs only on the south side of Montezuma County's Sleeping Ute Mountain, where gravel ridges formed by erosion of the mountain have buried the shale badlands flanking the mountain. These gravel terraces, called pediments by geologists, support a somewhat different assemblage of plants compared to the shale badlands a short distance away.

Neither of these milkvetches may flower in a dry year, but they will persist mainly underground, waiting for adequate rainfall. John Anderson and Mark Porter, who described and named *Astragalus tortipes*, noted that only one plant with fruit was located in 1985, and that few bloomed in 1990. When rains do permit good growth, twisted milkvetch is tall and robust, up to 3 feet tall with lemon yellow flowers. Cronquist milkvetch is a wiry-stemmed, spreading plant with dull pink-purple flowers. Stems and foliage bear scale-like hairs that are both filiform and flattened.

The sites of these milkvetches in the southwestern corner of Colorado are threatened by road construction, oil and gas exploration, and an irrigation system for the Ute Mountain Ute Indian Reservation. In fact, the discovery of twisted milkvetch came as a result of plant inventories conducted in association with work on seismic lines for oil and gas exploration and the irrigation canal system.

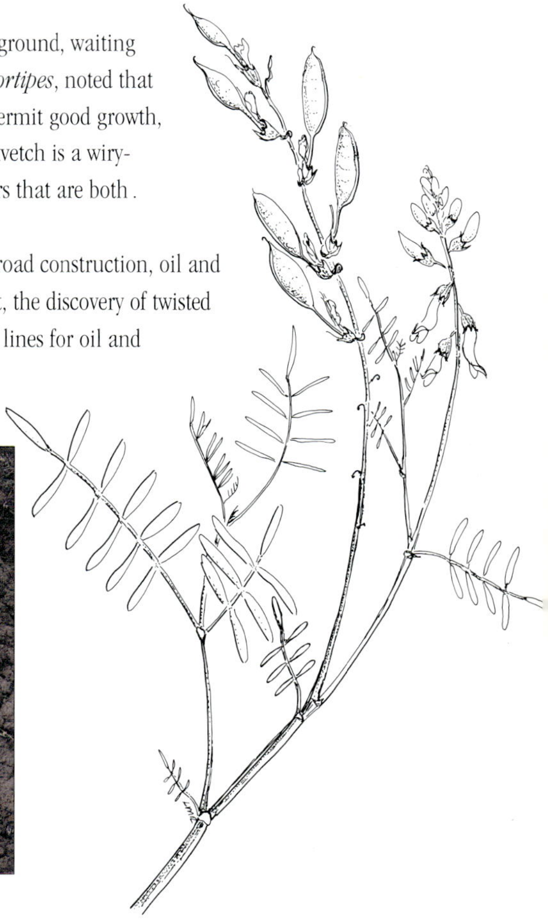

Heil

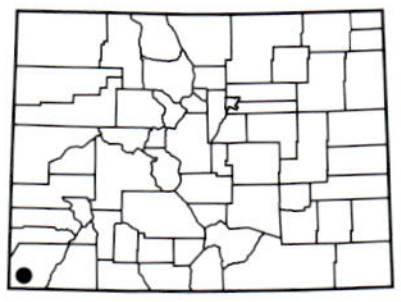

Astragalus tortipes

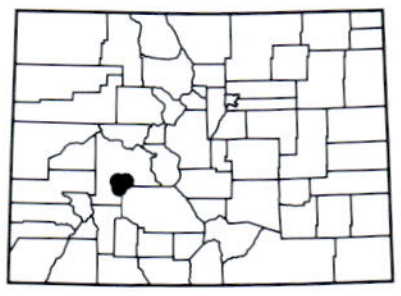

Astragalus microcymbus

C. Dawson

SKIFF MILKVETCH (Pea Family)
Astragalus microcymbus Barneby
GUNNISON BASIN MILKVETCH (Pea Family)
Astragalus anisus M.E. Jones

Late in the evening of July 20, 1945, at the end of a long day in the field, Rupert Barneby, an expert on the genus *Astragalus*, found some plants of a milkvetch growing by a roadside near Gunnison. Attempting to discover more plants for further study, Dr. Barneby and University of Colorado botanist William A. Weber repeatedly searched the site, but the milkvetch was not seen again until 1955, when Dr. Weber located eleven plants. Both botanists were perplexed by the plants and finally concluded that skiff milkvetch must be a "waif" at that locality, somehow introduced from afar. Dr. Barneby speculated that the plants might have originated in Mexico, where this species' closest relatives were best developed.

Joseph Barrell, a botanist long associated with the Rocky Mountain Biological Laboratory at Gothic, Colorado, and a specialist in the botany of the Gunnison Basin, disagreed with Barneby and Weber's waif theory, but was unable to find any other populations of the plants. In 1966, Dr. Barrell and his assistant were sitting down to lunch in a grove of cottonwoods southwest of Gunnison. There, they saw an odd-looking *Astragalus* against a low cliff. "Wouldn't it be a joke," said Dr. Barrell, "if that was *Astragalus microcymbus*?" It was! Skiff milkvetch since has been found in a few nearby locations well away from roads and buildings. To date, it is known to occur only in and near the South Beaver Creek drainage.

Skiff milkvetch is a rather tall, freely branching plant with many ¼-inch-long white flowers. The equally small, pendulous fruits reminded Dr. Barneby of an inverted skiff. The scientific name translates as little boat. Gunnison Basin milkvetch is a low-growing cushion plant with grayish foliage and very large, inflated, hairy, egg-shaped pods. The gray color of the plant is caused by the presence of numerous short-stalked, T-shaped (dolabriform) hairs.

Like skiff milkvetch, Gunnison Basin milkvetch, *Astragalus anisus*, was a mysterious plant for many years. When described by Marcus Jones in 1893, it was reported to be from Pueblo. Despite repeated searches, no plants have ever been found in that area. In the Gunnison Basin, it appears that botanists do their best work at lunchtime. In 1949, Dr. Weber was eating lunch on a rocky hillside east of Gunnison when he spotted this milkvetch. *Astragalus anisus* subsequently has been found occasionally on the sagebrush floor of the Gunnison Basin.

Jennings

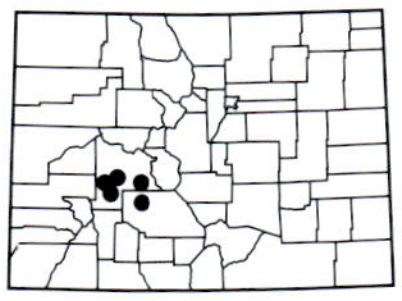

Astragalus anisus

Short-flowered Cryptantha, Uinta Basin Cryptantha (Borage Family)

Cryptantha breviflora (Osterhout) Payson

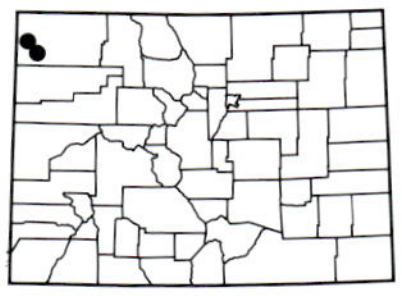

George Osterhout was an amateur botanist who lived at Windsor in Weld County, Colorado, where he owned a lumberyard. He began collecting plants in 1893, and by his death in 1937 had amassed a personal herbarium of 20,000 specimens. His specimens now reside at the University of Wyoming (Rocky Mountain Herbarium). Osterhout was an astute observer who found and described many plants still considered rare.

In 1925, he found an unusual cryptantha near the dinosaur quarry at Dinosaur National Monument, just across the state line in Utah. The plant still occurs near the dinosaur quarry, and recent research shows it to be at scattered locations in the Uinta Basin of Utah. In Colorado, there are two known sites in extreme western Moffat County.

Short-flowered cryptantha, *Cryptantha breviflora*, grows on steep clay barrens with sparse vegetative cover in the salt desert shrub, sagebrush, and pinyon-juniper zones. This plant could have been discussed just as well in the barrens chapter or the woodlands chapter of this book, depending upon the individual's perspective of the habitat. Is this cryptantha growing in a bare spot in the sage or a sage area in the barrens? Is it in a lightly wooded spot in the sage or is it a bare spot in the woods? Such uncertainty illustrates the problems of classifying plant communities, where boundaries are seldom distinct and plants occur in a mosaic of habitats.

This is a distinctive species of cryptantha, distinguished by its mound-forming habit, silky and glistening leaves, and flowers that are relatively large compared to other species in this genus.

Adobe Penstemon (Figwort Family)

Penstemon retrorsus Payson ex Pennell

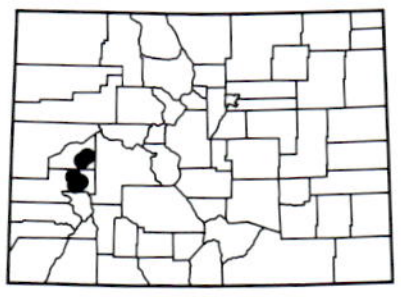

The leaves and stems of *Penstemon retrorsus* are ash colored because of coarse, dense hairs which point backward toward the base of the leaves or the stems. This characteristic is the basis for the name *retrorsus*, meaning backward. The plant has a woody rootstock that gives rise to a mat of stems and leaves. The flowering stems stand erect from the basal leaf rosette and provide an attractive display of small bluish purple flowers. Known from north-facing slopes on adobe hills derived from Mancos Shale in Montrose and Delta counties, plants of adobe penstemon sometimes are hidden among the junipers, yuccas, saltbush, and sagebrush.

Recreational use of the adobe clay hills by off-road vehicles and vegetational modifications to provide range for livestock have been the greatest threats to adobe penstemon. On the other hand, sheep do not eat *Penstemon retrorsus*, preferring other species such as saltbush. This may actually decrease competition, increasing the abundance of the penstemon. Emerging threats to the plant are the land subdivision and residential construction that accompany the area's increasing popularity as a retirement site.

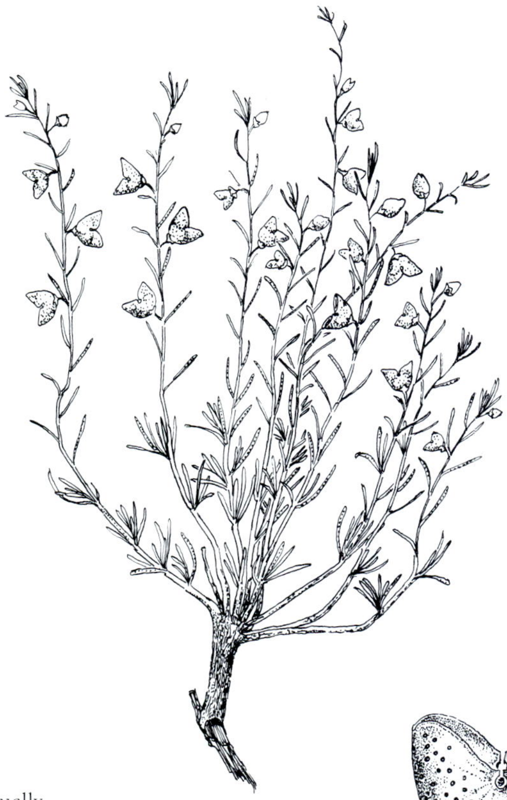

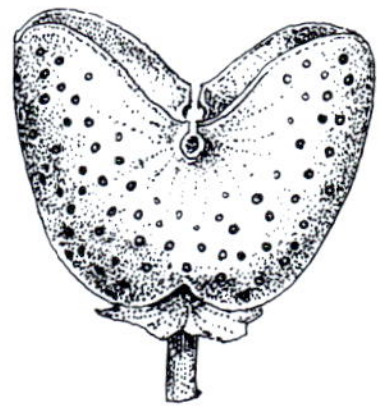

RUDA DEL MONTE, TEXAS DESERT RUE, DUTCHMAN'S BREECHES (Citrus Family)

Thamnosma texana (A. Gray) Torrey

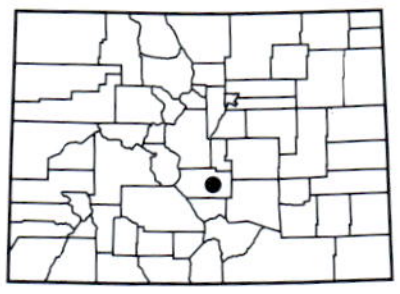

Ruda del monte is a loose, sprawling plant, usually 6 to 12 inches tall and found on rocky slopes and in open grassy areas. One might not guess that it is a member of the citrus family until leaves and stems are crushed and the characteristic lemony citrus odor is noticed. The leaves are covered with aromatic glands, and herbarium specimens retain their scent for decades. The small yellow flowers have four petals and four sepals. The fruit, an inflated two-lobed capsule, resembles a pair of pantaloons, the source of one of the common names.

As is currently understood, the range of *Thamnosma texana* is southern Arizona, southern and eastern New Mexico, central and western Texas, and the Mexican state of Chihuahua. The plant has been collected only twice in Colorado, both times from Cañon City. Specimens were collected in 1875 by T.S. Brandegee and again in 1877 by visiting British botanist J.D. Hooker, who undoubtedly was shown the plants by Brandegee. Brandegee stated that the collection site was "Soda Spring Ledge." William A. Weber, with help from Cara Fisher of the Cañon City Public Library, traced the location to what is now the southwest corner of the state penitentiary. However, Soda Spring was destroyed by highway construction on U.S. Highway 50.

Whether ruda del monte is native to Colorado is an unanswered question. The Arkansas Valley site is a considerable distance from the nearest current locations in the Texas Panhandle and eastern New Mexico. Legend says that Ute medicine men used the springwaters for curative treatment. Perhaps Native Americans brought the plant in, planted it, and encouraged it as a medicinal. On the other hand, there are a number of other plant species that occur in the Pueblo-Cañon City area, some distance from their main ranges in Texas and New Mexico.

Brandegee himself provided some clues as to where to look for the species. In 1874, just a year prior to collecting the *Thamnosma*, he collected windmills, *Allionia incarnata*, and groundsel tree, *Baccharis wrightii*, at Soda Spring Ledge. Both these species are known from southeastern Colorado in the Mesa de Maya region of Las Animas County. Habitat for ruda del monte may still exist there.

Jennings

HARRINGTON PENSTEMON (Figwort Family)

Penstemon harringtonii Penland

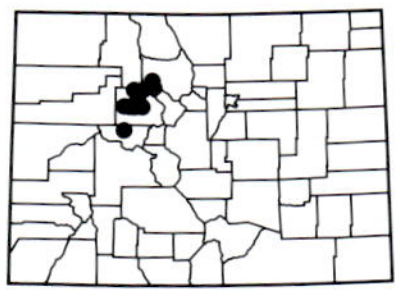

In the upper Colorado and lower Eagle river valleys, mostly in Grand and Eagle counties on slopes covered with pinyon pine, juniper and sagebrush, there are scattered populations of *Penstemon harringtonii*. This big, showy penstemon that grows from 1 to 2 ½ feet tall produces inch-long, deep blue to pinkish lilac flowers on about half the stem length. Two of the stamens extend out of the throat of the flower and two do not, important distinguishing features of this species.

Discovered in 1951, the plant was named for H.D. Harrington, long-time professor at Colorado State University and author of *Manual of the Plants of Colorado*, the most recent flora covering the entire state. C.W.T. Penland, a *Penstemon* specialist who taught at The Colorado College in Colorado Springs, named the species.

Some populations of Harrington penstemon may be threatened by grazing or land development for homes and ski resorts.

SMALL-FLOWERED PENSTEMON (Figwort Family)

Penstemon parviflorus Pennell

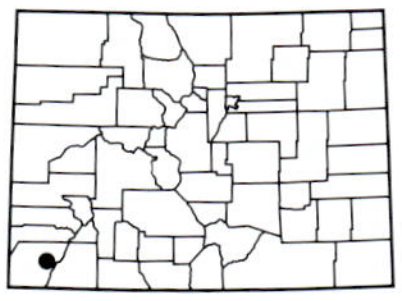

This book is full of articles about rare plants that were seen by some early-day botanist and not viewed again until nearly 100 years later. The happy result was that such plants, although rare, had managed to hold their own against decades of land misuse. Unfortunately, this is not the case with small-flowered penstemon, *Penstemon parviflorus*. In July of 1890, Alice Eastwood collected an unusual penstemon somewhere near Mancos, Montezuma County. *Penstemon parviflorus* was described from that single collection. Regrettably, it has not been seen since and is presumed extinct.

The type specimen plant has erect stems, leaves in whorls of three, and blue or purple flowers ½ inch long.

UINTA GILIA (Phlox Family)

Gilia stenothyrsa Gray

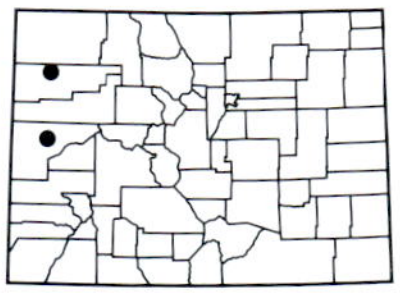

In the mid-1840s, on one of his expeditions to the West, General John C. Frémont found the Uinta gilia, *Gilia stenothyrsa*, near what is now Duchesne, Utah. His specimen eventually found its way to Harvard University, where it was recognized and described as a new species by Asa Gray, probably the best-known American botanist of the nineteenth century. Uinta gilia still is seldom encountered. Restricted to northeastern Utah and northwestern Colorado, it usually is found with pinyon and juniper but occasionally, it is found with with sagebrush or greasewood.

Gilia stenothyrsa is a biennial, meaning it takes two years for the plant to mature, flower, produce fruit, and die. From a rosette of basal leaves, the stout flowering stem can reach a height of more than 2 feet. The flowers, which appear in May and June, are white to pale purple with the anthers extending well past the tips of the petals. The entire plant is covered with minute, short-stalked, glandular hairs.

Jennings

NORTH PARK PHACELIA (Waterleaf Family)

Phacelia formosula Osterhout

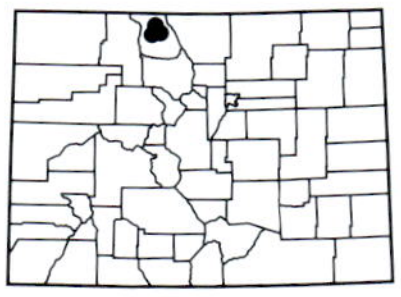

Jackson County encompasses the large mountain valley known as North Park. The North Platte River arises there and flows northward into Wyoming before turning eastward to join the South Platte in Nebraska. This is a land of sharp contrasts. Because there are a number of large streams, parts of the valley are quite wet and many moisture-loving plants are found there. But away from the rivers the land dries out very quickly, even allowing the formation of sand dunes. On easily eroding sandstone outcrops of the Coalmont Formation, and only in North Park, the North Park phacelia is found.

Discovered near Walden by George Osterhout in 1918, *Phacelia formosula* was known exclusively from that site for many years. In 1980, only 117 plants were found there. After intensive searching in the area, however, several other populations have been discovered. North Park phacelia is listed as an Endangered Species by the U.S. Fish and Wildlife Service, which has the responsibility of implementing the national Endangered Species Act.

In July and August, *Phacelia formosula* produces a handsome display of numerous purple flowers. It apparently does not compete well with other vegetation, instead growing where little else will grow. The Colorado Natural Areas Program, Bureau of Land Management, and U.S. Fish and Wildlife Service are cooperating in a monitoring study and seek to assure the long-term preservation of this plant.

Jennings

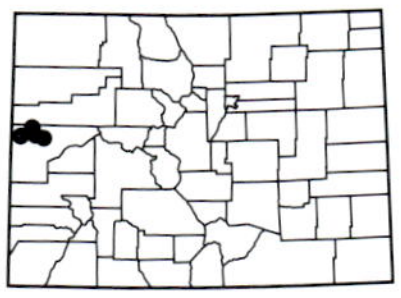

Grand Wild Buckwheat

Eriogonum contortum Small ex Rydberg

Buckwheats (Buckwheat Family)

It often seems that rare plants are known from only one specimen which languishes either unknown or misidentified in some herbarium collection. Eventually, a botanical specialist sees the specimen and is struck by its differences from related species. This was the case with clay-loving buckwheat, *Eriogonum pelinophilum*, known from a single specimen collected by Harold Gentry in 1958. Upon studying the specimen, James Reveal set out to find more plants but was unsuccessful until 1972. In 1984 and 1985, the plant was found in a few more places in Montrose and Delta counties, in sparsely vegetated shrublands and on the barren adobe clay soils derived from the Mancos Shale. The plant is listed by the U.S. Fish and Wildlife Service under the Endangered Species Act as Endangered.

Eriogonum pelinophilum is a small shrub, about 5 inches tall and up to a foot wide, arising from a woody taproot. It has numerous woody branches from which bark hangs off in loose strips. Leaves are scattered the length of the stem, rather closely spaced and congested. The margins of the leaves are revolute, or rolled back, nearly covering the lower surface and resembling tiny scrolls. The plant blooms in June and July, producing many miniature flowers that are white to cream colored with reddish brown midribs.

Like the milkvetches and penstemons, wild buckwheats tend to occupy specialized habitats of limited distribution. There are a number of rare species found on the Colorado Plateau and in the Great Basin. Comb Wash wild buckwheat, *Eriogonum clavellatum*, and Grand wild buckwheat, *Eriogonum contortum*, along with clay-loving wild buckwheat discussed above are examples of such species. Botanists use the term endemic to describe plants of this sort, which occur only within a specific, delimited area or region. Endemics may be rare or common, but when an endemic plant is rare there must be considerable concern for potential loss of its limited habitat. Endemic rare plants may be contrasted with those exemplifying other types of rarity, such as those that are widespread geographically but nowhere common.

Comb Wash wild buckwheat, *E. clavellatum*, can be found on shales in San Juan County, Utah, and adjacent Montezuma County, Colorado, often with the equally rare *Astragalus cronquistii* and *Astragalus tortipes*. *Eriogonum clavellatum* is a low shrub, less than 8 inches tall with very narrow leaves less than 1/8 inch wide. Flowers are white, in some years making a showy display.

Eriogonum contortum is limited to Grand County, Utah, and adjacent Mesa County, Colorado, mostly on the sparsely vegetated shales north of the Colorado River. Also a low shrub, it has leaves recurved like miniature scrolls. Its flowers are yellow.

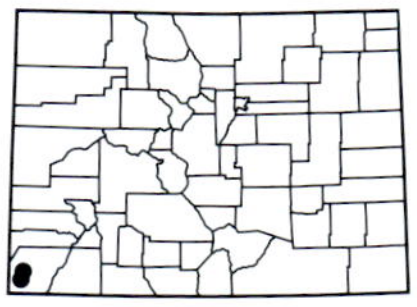

Comb Wash Wild Buckwheat
Eriogonum clavellatum Small

Yeatts

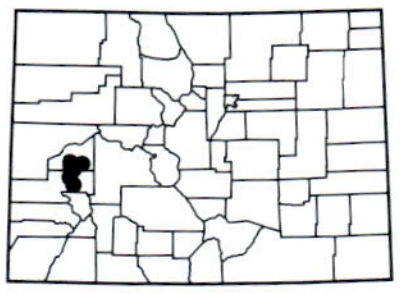

Clay-loving Wild Buckwheat
Eriogonum pelinophilum Reveal

Steinkamp

PARADOX VALLEY CRYPTANTHA (Borage Family)

Cryptantha paradoxa (A. Nelson) Payson

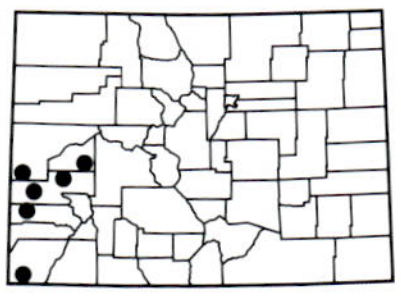

In 1909, eighteen-year-old Ernest P. Walker and his parents moved to the Paradox Valley in extreme western Montrose County. This was then, as now, a remote and seldom-visited portion of the state. Walker began making plant collections, which eventually were given to Aven Nelson, his teacher at the University of Wyoming. Walker's interests soon turned to zoology, and later in his career he was assistant director of the National Zoological Park in Washington, D.C.

In 1912, Walker collected a small cryptantha, which Nelson determined was an undescribed species. Presumably, the plant was named *Cryptantha paradoxa* for the location of its discovery, not because of any self-contradictory features of the plant. But paradoxical it is, because it grows on soils with a high gypsum content or on outcrops of pure gypsum, places "where plants don't grow." Gypsum is a naturally occurring mineral, hydrous calcium sulfate, produced as water evaporated from ancient saline lakes. The plant is further paradoxical in that it has been found on heavy clay soils that are not particularly high in gypsum.

In Utah, *Cryptantha paradoxa* is reported from soils derived from sandstone. Some clays are high in calcium and some sandstones are cemented with calcium carbonate, and the presence of high concentrations of calcium may be the uniting feature for the diverse substrates on which this plant is found. On the other hand, the plant may be well adapted to aridity, or it may be unable to compete with other species in more mesic situations but can tolerate sites others cannot. Clearly, more research is needed to determine whether plants growing outside the Paradox Basin on rocks other than gypsum might represent one or more undescribed species.

Paradox Valley cryptantha is a low plant, usually less than 4 inches tall with very hairy leaves and attractive, trumpet-shaped white flowers with yellow prominences near the center.

CUPPED PENSTEMON, MIDDLE PARK PENSTEMON (Figwort Family)

Penstemon cyathophorus Rydberg

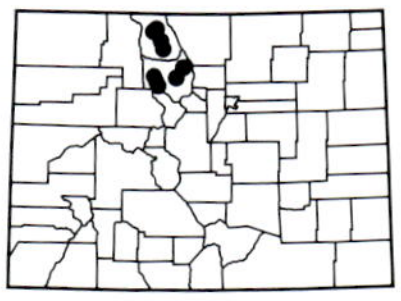

Like the other penstemons described in the accompanying articles, cupped penstemon, *Penstemon cyathophorus*, has a limited range and specific requirements. First found in 1901 in North Park, it also occurs in a few scattered populations west of Granby and in Sweetwater County, Wyoming. In all localities, it is found in small patches among sagebrush.

The common name, cupped penstemon, comes from an allusion to the broad bracts that subtend the pink flowers; *cyathophorus* means cup-bearing. The most notable features of the plant, however, are the four stamens that are exserted, meaning they extend beyond the mouth of the flower. The blue-flowered Harrington's penstemon, *Penstemon harringtonii*, has two similar exserted stamens. The two species have been reported together near Kremmling, but in general, the range of cupped penstemon is to the north and east of Harrington's penstemon.

Jennings

Steinkamp

Steinkamp

Osterhout Milkvetch (Pea Family)

Astragalus osterhoutii M.E. Jones

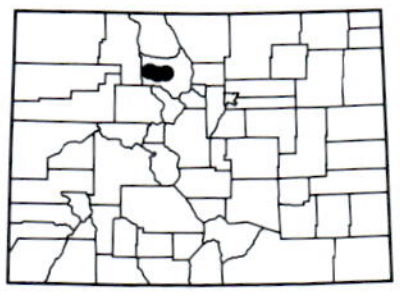

The common name milkvetch alludes to the ability of some plants in the genus *Astragalus* to enhance cows' milk production. Other plants in this genus are called locoweed or crazyweed because livestock that feed upon them react in unpredictable fashion and are said to go *loco*, Spanish for crazy. Often, plants that grow in high-selenium soils concentrate the element in their tissues and produce a characteristic unpleasant odor. When these plants are grazed, the ailment known as blind staggers or alkalai disease may result. Osterhout's milkvetch, found in the Kremmling area, grows on high-selenium soils derived from the Niobrara and Troublesome formations and possesses the characteristic selenium smell.

Osterhout's milkvetch is restricted to Grand County, where four known populations total perhaps 3,000 individual plants. In early summer, it produces an attractive display of many large, snowy white flowers followed by pendulous 2-inch-long pods typical of the genus.

A significant part of the known range and population of *Astragalus osterhoutii* was lost when a new reservoir on Muddy Creek was filled for the first time in 1995. Other threats include trampling by grazing animals.

Under the provisions of the Endangered Species Act, Osterhout's milkvetch is listed as Endangered.

USFS/Johnston

UINTA BASIN HOOKLESS CACTUS (Cactus Family)

Sclerocactus glaucus (K. Schumann) L. Benson

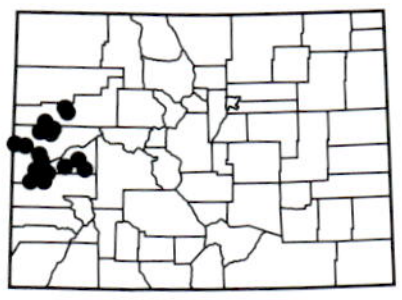

The gravelly soil of hills and mesas in the Colorado and Gunnison river valleys is the home of the Uinta Basin hookless cactus, a small ball cactus about 2 or 3 inches in diameter. Even when in bloom, the plants blend with the rocks and gravels in these hot, dry sites and can only be found after diligent search. The reward is worth the effort, for the flowers ranging from pink to magenta are most attractive. After blooming, the cactus shrinks beneath the surface and usually is impossible to find.

Sclerocactus glaucus was discovered in Colorado in 1892 by the German botanist C.A. Purpus. Originally described as an *Echinocactus*, the plant later was placed in *Sclerocactus* by Lyman Benson, a specialist in the cactus family.

Also found in western Colorado is the similar but more common species *Sclerocactus whipplei*, which usually is larger and has hooked spines. The lack of these hooked spines is a distinguishing characteristic of *Sclerocactus glaucus*. Hybridization between the two species may occur as both straight and hooked spines are found on some plants. Uinta Basin hookless cactus is protected by the federal Endangered Species Act, under which it is listed as Threatened. Mineral development and collectors of unusual cacti pose the primary threats to the species.

COLORADO DESERT-PARSLEY (Parsley Family)

Lomatium concinnum (Osterhout) Mathias

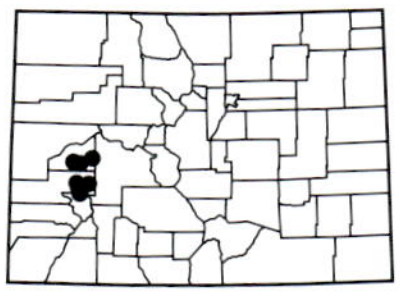

In 1911, George Osterhout found an interesting parsley in Delta County near Paonia. He described his find as *Cogswellia concinna*, the name *concinna* meaning neat or elegant. In the 1930s, Mildred Mathias, an expert on the parsley family, transferred the species to *Lomatium*. Osterhout wasn't the first to see the plant as it had been collected by others in 1893 and 1901. After Osterhout's 1911 collection, however, the species was not collected again for 41 years until William A. Weber located it in Montrose County. In recent years, a few more populations have been found in Delta and Ouray counties.

The Colorado desert-parsley, endemic to the barren adobe hills derived from the decomposition of the Mancos Shale, grows in salt-bush communities. *Lomatium concinnum*, which displays shiny green leaves and yellow flowers, sometimes occurs with two other rare shrubland plants, *Penstemon retrorsus* and *Eriogonum pelinophilum*.

Jennings

PENLAND BEARDTONGUE (Figwort Family)

Penstemon penlandii W.A. Weber

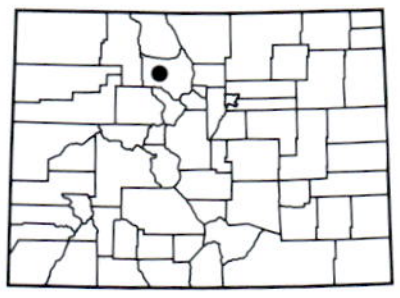

Near Kremmling are areas underlain by the Troublesome Formation, which is composed of shales that weather into barren clay slopes. These alkaline clays contain selenium, which together with the peculiar mineralogical properties of clay soils effectively excludes almost all vegetation. But the clay knolls are home to a tough little penstemon, the Penland beardtongue, *Penstemon penlandii*.

Located in 1986 during a survey related to the relocation of a power line, the penstemon was brought to the attention of William A. Weber of the University of Colorado. Dr. Weber visited the site and noted that the most luxuriant plants were those growing from seed which had been washed downslope by torrential rains and had germinated in deep runoff channels in the shade produced by the cut banks. *Penstemon penlandii* develops root structures that enable it to hold fast to the underlying shales so it will not be dislodged by subsequent torrents, a feature common to many plants of barren lands. Where erosion has mixed the seleniferous clays with less toxic materials the penstemon disappears and sagebrush takes over. The one known population of *Penstemon penlandii* is associated with *Astragalus osterhoutii*, an almost equally rare plant described elsewhere in this chapter.

Plants of Penland beardtongue grow up to 10 inches tall. Stems are laden with small flowers that are blue violet when fresh, fading to pinkish blue. The basal leaves are narrow and linear, but inconspicuous.

Penland beardtongue is listed as Endangered under the federal Endangered Species Act.

HAIRY EASTER DAISY (Daisy Family)

Townsendia strigosa Nuttall

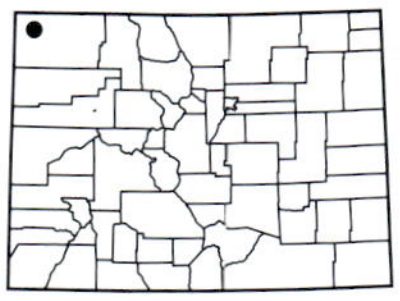

Thomas Nuttall observed and collected *Townsendia strigosa* in 1834 while traveling across central Wyoming. Even though this species has been known to science for more than 150 years, it has been found only sparingly in southwestern Wyoming, northeastern Utah, and in Moffat County, Colorado, where it was discovered in 1988 by J. Scott Peterson and Dieter Wilken. Hairy Easter daisy can be found on shaly, sandy, or alkaline clay substrates in desert shrub, sage-brush, or pinyon-juniper habitats.

Plants in the genus *Townsendia* are called Easter daisy because some species bloom early in the spring, particularly around Easter. Some species are known to bloom as early as February. Hairy Easter daisy is a spring bloomer (May and June), but not as early as some. It is a winter annual or biennial with a large taproot developed the first season. The second year, the dead leaves of the first year's basal rosette clothe the base of the stem. As the common name implies, plants are covered with long hairs. These, however, are not so dense as to obscure a reddish pigmentation of the stem. Ray flowers are white and disk flowers are yellow.

Wetlands

Wetlands differ from the other habitats described in this book in two ways. First, they occur as inclusions in the broader habitats of woodlands, forests, barrens, shrublands, alpine tundra, grasslands, and canyons. Second, wetlands are the only habitat type afforded general federal protection, whether they occur on private or public lands. In Colorado, wetlands most frequently exist either along streams, pond margins, and lakeshores, or in areas with a high groundwater table. Rare plants can be found in wetlands at any elevation. For example, *Spiranthes diluvialis* occurs along streams east of the Front Range. Slender spiderflower, *Cleome multicaulis*, is limited to San Luis Valley wetlands. Altai cottongrass, *Eriophorum altaicum*, is present only in the highest wetlands of southwestern Colorado. Rare subalpine wetland plants in Colorado include three species of willow (*Salix* sp.); pale blue-eyed grass, *Sisyrinchium pallidum*; and Greenland primrose, *Primula egaliksensis*.

Because the federal Clean Water Act prohibits the filling of wetlands without a permit granted by the U.S. Army Corps of Engineers and reviewed by the Environmental Protection Agency, a legal definition of a wetland has been established. According to the government, wetlands are ". . . those areas that are inundated or saturated by surface or ground water at a frequency and duration sufficient to support, and that under normal circumstances do support, a prevalence of vegetation typically adapted for life in saturated soil conditions. Wetlands generally include swamps, marshes, bogs, and similar areas."

Although wetlands are found throughout Colorado, they are distributed sparingly, comprising about 3 percent of the state's area. Wetlands really are microhabitats within a larger landscape unit. For example, a small spring, seep, cattail marsh, or drainageway may be surrounded by a grassland or other habitat type. Relative to the habitat around it, the most important feature of a wetland is the greater abundance of water, an extremely important resource in semiarid Colorado. Within the wetland, microscale changes in elevation can bring about large changes in plant communities. An example of this is the hummock topography found in subalpine wetlands.

The wetland distribution we see today is the result of earlier geologic events and climatic changes. For example, mountain glaciers are responsible for forming many of the subalpine wetlands. The dominant species on the drier hummocks are different from species dominating the valleys, even though the difference in elevation is no more than 6 inches.

The broad, U-shaped valleys formed by glaciation are more favorable for wetland formation than the steep-walled, V-shaped canyons of the lower mountains and foothills. Other glacial features such as terminal moraines damming a valley or partially filled ancient glacial lakes help form wetlands in the mountains. On the plains, depressions that probably were lakes in wetter times now are marshes.

The maintenance or regeneration of some wetland communities is dependent upon a cycle of natural disturbance. The cycle of flooding, sediment deposition, uprooting of old communities, and the formation of unvegetated sites ripe for colonization is typical for the riparian wetlands that are found along the larger rivers. Natural stream meandering, the formation of oxbow lakes, and the sedimentation of old channels also affect wetland formation and maintenance. Subalpine and montane wetlands have a cycle of natural disturbance that involves the beaver. Damming, abandonment, and subsequent filling or breaching of beaver dams contribute to the dynamics of these wetlands, as does the harvesting of trees and shrubs by the beaver.

Early settlers began to change the distribution of wetlands immediately upon their initiation of agriculture. Low, poorly drained marshes were drained completely or filled. Shrub-dominated wetlands were cleared and converted to hay meadows. As water rights were bought and sold, water manipulation took place. Water was stored for irrigation and slowly released, causing ephemeral streams to be perennial while water diversion projects had the reverse effect. Flood control dams stopped seasonal floods. Although modern water development continues to modify wetlands and controversial water projects still are being built, the Army Corps of Engineers-Environmental Protection Agency permitting process mentioned earlier includes an important mitigation component. There have been wetland restoration, enhancement, and development projects in the state. Wetland restoration projects include the Chatfield Arboretum, Boulder Creek through the City of Boulder, and the Blue River near Breckenridge. However, many wetlands cannot be restored once damaged or abused. Some high-altitude wetlands, for example, date from the last glaciation and cannot be replaced if lost. The wetland destruction that currently is permitted through harvest and sale of so-called Colorado mountain peat is a sad example of this.

Delicate Sedge, Bristle-stalk Sedge (Sedge Family)

Carex leptalea Wahlenberg

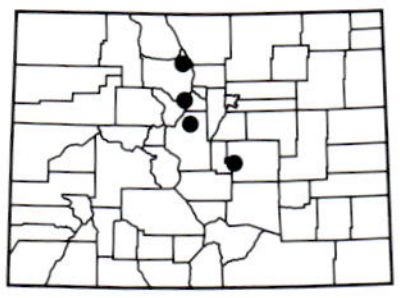

The word sedge is derived from Middle English and Teutonic words that mean saw as many sedges have sharp-edged leaves. The word has been in usage for more than a thousand years. The Oxford English Dictionary indicates that not only the true sedges but other grass-like wetland plants probably were included under the definition. References of the usage are cited to A.D. 725, but a most interesting one dates from 1440: "Segge, star of the fenne, *Carix*." Although the spelling has changed since 1440, the meaning still is the same.

The word delicate aptly describes *Carex leptalea*. The culms (stems) and leaves are nearly as thin as thread. Growing almost a foot tall in Colorado, this sedge seems to defy gravity when standing erect.

Primarily a Canadian species, *Carex leptalea* is scattered sparingly in the Rocky Mountains. It is known from the Uinta Mountains of Utah, Yellowstone National Park in Wyoming, and the Black Hills of South Dakota. Prime habitat is in moist canyons or rich fens. In Colorado, delicate or bristle-stalk sedge first was collected by pioneer botanists Elihu Hall and J.P. Harbour in 1862 in South Park. Since that time, it has been found only three times — in El Paso County in 1956, and in Clear Creek and Grand counties in 1994.

Gray Sedge (Sedge Family)

Carex livida (Wahlenberg) Willdenow

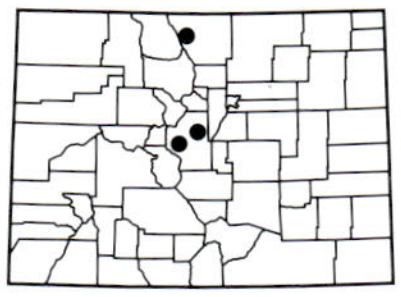

Gray sedge, *Carex livida*, has a whitish, waxy coating on the stems, providing the basis for both the common and scientific names. Circumboreal, gray sedge occurs in swampy or boggy ground in the far northern latitudes of North America, Europe, and Asia. The nearest populations to Colorado occur in Yellowstone National Park. Good habitat for gray sedge is rare in Colorado, but the plant is present in a few rich fens where it grows with other rare sedges on a thick mat of peat. Colorado occurrences of this plant are in Boston Peak Fen in Larimer County and in High Creek Fen and East Lost Park, both located in Park County.

In Colorado, sedges are low plants found primarily in wet places in the mountains. Sedges usually are unpalatable to livestock and few are of any importance to commerce. However, papyrus, famed from antiquity as the raw material for paper, is a member of the sedge family.

Slender Cottongrass (Sedge Family)

Eriophorum gracile W.D.J. Koch in Roth

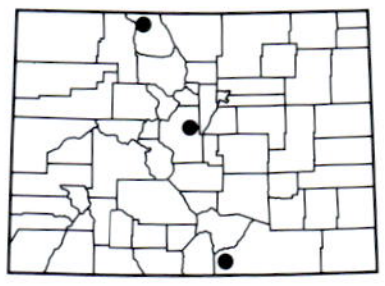

There are a few specimens of *Eriophorum gracile* from the nutrient-rich, peaty wetlands of North Park and Lost Park, but few from elsewhere in the state. Otherwise, like many other high-altitude wetland plants, slender cottongrass occurs only much farther north. It can be distinguished from the other two Colorado cottongrasses by its short bristles, which result in a much smaller "cottonball." The tendency to form large clumps and the characteristic reddish color of the leaf tips make the plant identifiable from a distance.

The scientific name of the genus, *Eriophorum*, is taken from two Greek words that mean wool-bearing, an allusion to the white, cottony heads. Other common names given to these plants are bogwool and cottonsedge, both reasonable considering the derivation of the scientific name, the habitat, and the fact that the plants are members of the sedge family, not grasses.

Many-headed Sedge (Sedge Family)

Carex synchnocephala Carey

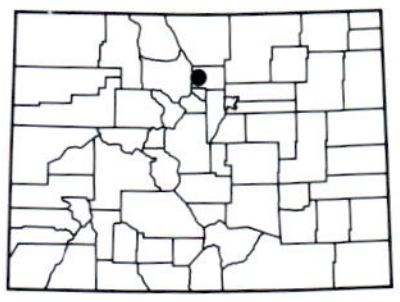

Few species of sedges, if any, are Colorado endemics, meaning limited in range solely to Colorado. Most are species of the boreal regions of North America, occurring across Canada and Alaska. Their ranges often extend southward in high mountain wetlands. Many-headed sedge, *Carex synchnocephala*, is such a species. It has been recorded only once in Colorado, on the shore of a subalpine lake in Boulder County. The nearest other populations are in Montana and North Dakota.

There are more than 100 *Carex* species found in Colorado. Most are small wetland plants, and all lack showy flowers. Plants are best identified late in the season when mature fruit is present, so sedges found early in the season frequently require a second trip to the site for conclusive identification. In addition, they often have deep, tough, tightly massed roots. As a result, sedges have been avoided by most amateur and even professional botanists, a real loss to botanical science. All the accompanying articles about high-altitude wetland plants have a similar theme: The species is known from a few wet spots in Colorado, but otherwise it is a species of the far north. Are these species really that rare in Colorado? We don't know for sure, and it is time to make the commitment to understand our wetlands before they all are converted to reservoirs.

Vasey Bulrush (Rush Family)

Juncus vaseyi Engelmann

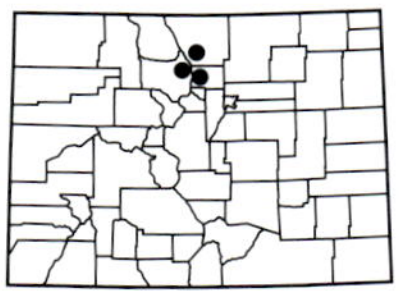

Plants of the rush family (Juncaceae) vegetatively resemble both the grasses and the sedges but actually are related more closely to the lily family. The flowers are small, dry, and scaly, but upon close examination can resemble miniature lily flowers. Vasey rush was named for the nineteenth-century botanist George Vasey, curator of the United States National Herbarium. The rush is differentiated from similar species by its leaves, which are nearly round in cross-section, and by its seeds, which include a membranous tail nearly as long as the seed body.

Rushes typically inhabit permanently moist localities, and Vasey rush is no exception, growing in wet meadows, margins of bogs, or depressions along sandy lakeshores. Uncommon throughout its range, which extends across southern Canada and the northern United States, *Juncus vaseyi* is found in the Rocky Mountains only in a few scattered localities in Idaho, Wyoming, and Colorado. The first specimens from Colorado were collected by Elihu Hall in 1862. Vasey rush has been seen only rarely since then, mostly near Grand Lake and in Rocky Mountain National Park. The plant was found at both localities in 1986, but these were the first reports since 1935. The disappearance of Vasey rush probably is related to the conversion of mountain wetlands to hay meadows.

Little Bulrush (Sedge Family)

Scirpus pumilus Vahl

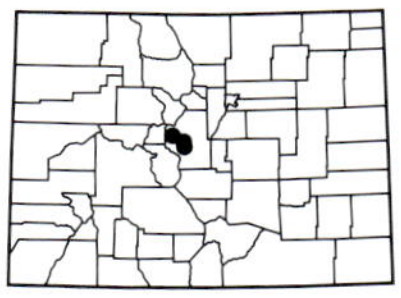

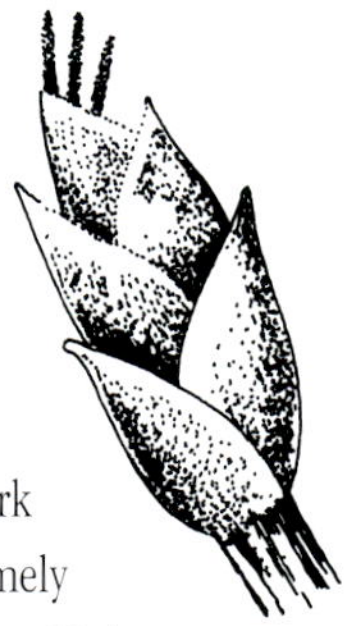

Named by Martin H. Vahl, professor of botany at Copenhagen University and a student of Linnaeus, the little bulrush is found in arctic regions of Eurasia and North America. Discovered in Colorado in 1862, little bulrush, *Scirpus pumilus*, was unreported for more than a century. In 1984, it was rediscovered high in the Mosquito Range of Park County, and a few other populations now have been found in the same area. Little bulrush is extremely inconspicuous, with only a few slender stems up to 4 inches tall. It grows in wet moss in willow carrs. Little bulrush produces a solitary flower cluster at the end of the stem.

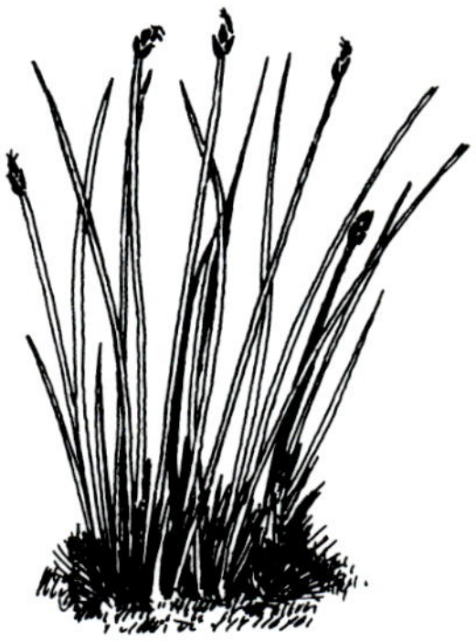

At its original locality high in the Colorado mountains, *Scirpus pumilus* grows on limestone. Because the original sedimentary layers, some of which were calcareous, mostly have eroded away from the tops of Colorado's high mountain ranges, the combination of high altitude and calcareous soils is most unusual here. Wherever the combination occurs, rare or unusual plants always are found. Unfortunately for these rare plants, many high-altitude limestone formations in Colorado are host rocks for major mineral deposits. Intensive development has occurred in these areas, beginning in the nineteenth century but continuing to the present, and much of this habitat has been lost. Peat mining in South Park also threatens populations found there in fens.

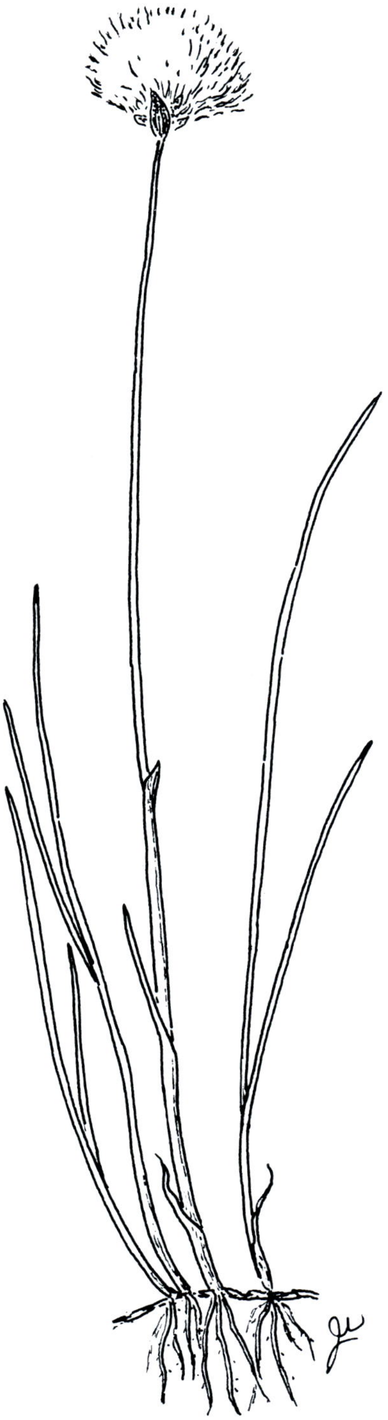

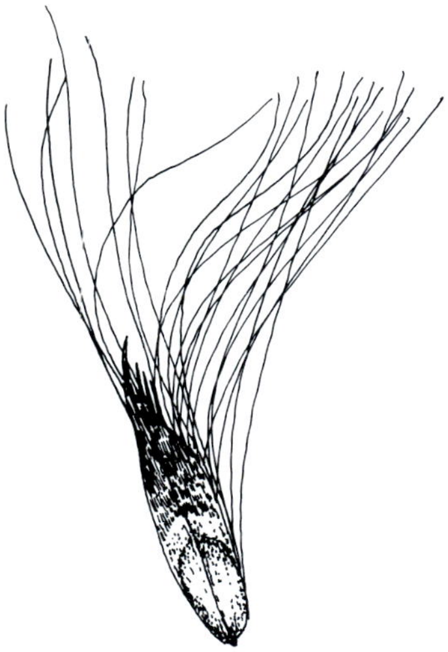

ALTAI COTTONGRASS, WHITE COTTONGRASS (Sedge Family)

Eriophorum altaicum Meinshausen var. ***neogaeum*** Raymond

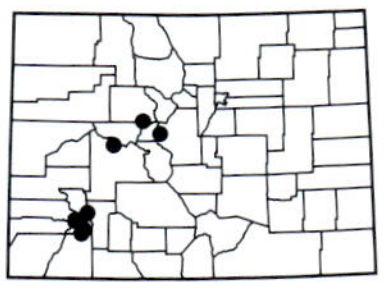

Altai cottongrass, *Eriophorum altaicum*, actually is a member of the sedge family, which includes plants that may look grass-like but are not true grasses. Cottongrass gets its name because its flower cluster resembles a powderpuff or ball of cotton. The petals and sepals, collectively called the perianth, consist of many long, white silky hairs. In Altai cottongrass, the "cotton" is white, whereas in a related species it is russet brown. In addition, the culm, which is the technical term in the grass and sedge families for what would be called the stem in other plants, lacks well-developed leaf blades. The whole plant looks like a cotton ball on a foot-long stick.

First found in the Altai Mountains of central Asia, *Eriophorum altaicum* now is known from alpine and arctic regions in much of the northern hemisphere. The North American plants differ slightly from Asian ones and have been assigned the varietal name *neogaeum*, meaning race "from the new world." Found in Colorado for the first time in 1934, Altai cottongrass since has been reported a few times from peat wetlands high in the southwestern mountains, especially near Silverton.

Jennings

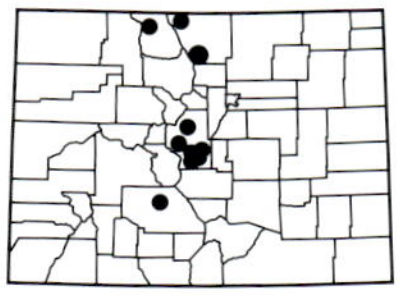

Pale Blue-eyed Grass (Iris Family)

Sisyrinchium pallidum Cholewa & Henderson

Jennings

In Colorado, some fen wetlands are formed when groundwater constantly flows upward to the surface. South of Fairplay in South Park is one such wetland known as High Creek Fen. An exceptionally diverse flora occurs there. High Creek Fen now is a preserve owned by The Nature Conservancy, a conservation organization that buys high-quality natural areas for preservation. During the past 8,000 years, a layer of peat formed at High Creek Fen, promoted by the cold, alkaline waters bubbling to the surface there. Such occurrences are unusual in Colorado, and they are greatly endangered by mining for mountain peat, draining for improved forage production, or efforts to increase runoff to reservoirs. A number of rare plants, including Greenland primrose, pale blue-eyed grass, and rayless groundsel, are present at High Creek Fen and similar fens in northern Colorado.

Greenland primrose, *Primula egaliksensis*, originally was found near the Inuit (Eskimo) village of Egaliko in Greenland early in the nineteenth century. Since then, it has been found occasionally in northern Canada and Alaska, and it was discovered in Wyoming in 1984. Although it first was collected in Colorado in 1936, the specimens were misidentified. Nearly 20 years elapsed before the plants were recognized as belonging to this species.

Often, *Primula egaliksensis* grows with the similar *Primula incana*, a more common species. In comparison with the latter, Greenland primrose is more slender and dainty, has smaller, nearly white flowers, and lacks a whitish coating on the backs of the leaves and on the stem.

Pale blue-eyed grass, *Sisyrinchium pallidum*, is not a grass at all, despite the grass-like appearance of its stem and leaves. Instead, it is a member of the iris family. The pale blue flower, a little smaller than a dime, is the clue that this can't be a grass, even though it doesn't look much like our familiar garden iris, either. Pale blue-eyed grass is found only in a few scattered locations in Saguache, Park, Jackson, and Larimer counties, Colorado, and Albany County, Wyoming. It inhabits fens and poorly drained meadows at about 7,000 to 10,000 feet elevation.

Rayless groundsel, *Packera pauciflora*, mostly is found in Canada, but there are a few scattered populations in northern Idaho and northern Wyoming. In Colorado, it is limited to the wetlands of South Park, where its attractive orange-yellow, button-shaped "flowers" make it particularly conspicuous.

Flowers in the daisy family actually are heads comprised of multiple flowers of two types: ray flowers, which would be called petals by many observers, and disk flowers, the smaller and densely packed flowers in the center of a typical daisy arrangement. In most species of the daisy family, both types of flowers are present, but either disk flowers or ray flowers sometimes may be lacking. In rayless groundsel, logically enough, there are no ray flowers.

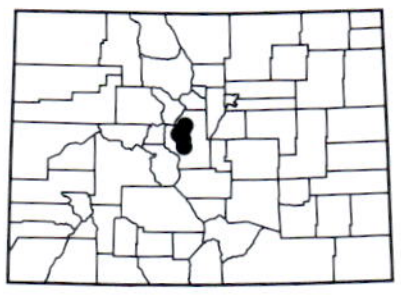

Greenland Primrose (Primrose Family)

Primula egaliksensis Wormskjold

Jennings

Jennings

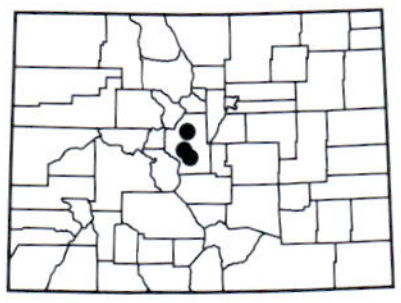

Rayless Groundsel (Daisy Family)

Packera pauciflora (Pursh) Löve & Löve

CNAP/Coles

Myrtle-leaf Willow, Low Blueberry Willow (Willow Family)

Salix myrtillifolia Andersson

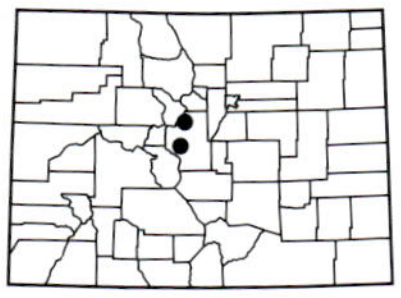

Fens and bogs are wetlands with peat soils comprised of partially decomposed, compacted organic matter. Peat often is dug and marketed as peat moss to be used as mulch, as a soil amendment, or cut and dried to be used as fuel. A fen is supported by surface water flowing laterally or groundwater flowing upward from springs or seeps. A bog is supported only by rainwater flowing vertically downward and usually is limited to very humid climates such as those found in northern Europe. A fen can become a bog when the peat mat builds up and is elevated above the level surrounding the fen surface.

In Colorado, peat fens have formed in a few high mountain valleys. Such areas are relicts of the moister, colder conditions that prevailed during the Pleistocene glacial periods. These fens, if undisturbed, harbor populations of plants now found primarily in northern Canada or Alaska. In a few places, particularly in South Park, peat has formed in sufficient quantity to be of economic interest and mined and sold as mountain peat. Once mined, a site cannot be reclaimed to the original vegetation because peat soil formation requires thousands of years. The Colorado Native Plant Society has joined other knowledgeable groups and individuals in urging that mountain peat **never** be purchased.

High Creek Fen near Fairplay, one of the largest such peat deposits, is especially rich in plant species typical of northern latitudes. In 1989, myrtle-leaf willow, *Salix myrtillifolia*, was discovered there by wetland ecologist David Cooper. Dr. Cooper reported that the willow was found on hummocks in the extremely rich center of the fen, which is fed by seeps of highly mineralized water, another unique feature of the South Park area. Other fens, though formed the same way as the South Park fens, are fed by waters low in mineral nutrients and the northern plants are not present.

Salix myrtillifolia is less than 3 feet tall, often trailing and rooting along the stem. Older branches are grayish brown but young twigs are greenish. Leaves are less than 2 inches long. The flowering structures, technically called aments, are up to 1 ½ inches long, appear with the leaves, and are borne on short, leafy branches. Found in much of Canada and Alaska, myrtle-leaf willow seldom is found in the contiguous forty-eight states.

Willow Reproduction

Members of the willow family (Salicaceae) bear male flowers and female flowers on different plants. Thus, a given plant is either male or female. Biologically speaking, plants having male and female elements on separate plants are said to be dioecióus (pronounced di-ee'-shus), a term derived from the Greek words *dis*, meaning two, and *oikos*, meaning house. Dioecious plants frequently are wind pollinated but in most species of willow, pollination is by insects. Flowers appear before the leaves are fully expanded. The familiar "pussy-willow" is the male flower cluster.

Steinkamp

Martin

Hoary Willow, Silver Willow, Sage-leaf Willow (Willow Family)

Salix candida Fluegge

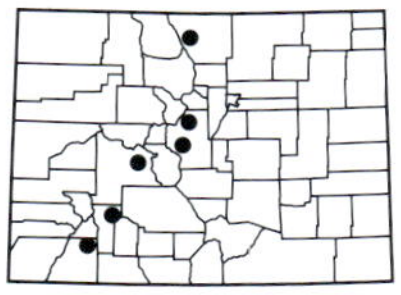

Like the other rare willows described in this chapter, the hoary willow is known from very few locations in Colorado, and the nearest other populations are far to the north. This inconspicuous willow shrub grows about 3 feet tall in wetlands at moderately high elevation. The stems are unbranched, except at the very end, where they produce a few short, leafy branchlets. New twigs and fruits are tomentose (densely covered with woolly or cottony hairs), giving the plant a frosted aspect. The undersides of the leaves are covered with short white hairs as well.

For thousands of years, people have used extracts from various species of willow to treat pain. The medicinal effect of willow is due to the presence of a chemical called salicin, a modified form of salicylic acid, and some related compounds. Although extract of willow may have relieved pain, it also caused very unpleasant intestinal distress. Today, we use a different derivative of salicylic acid, acetylsalicylic acid, as the active ingredient in aspirin.

Autumn Willow (Willow Family)

Salix serissima (Bailey) Fernald

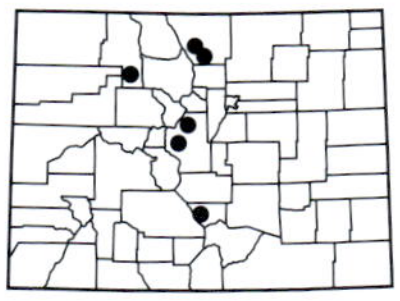

Thanks to intensive study of the willows by several workers in Colorado, the autumn willow now is known in our state from several locations. The original Colorado site, discovered by F.W. Hunnewell in 1923, is at 9,000 feet near Longs Peak. The nearest populations outside Colorado are in Montana, and the willow's range extends north and east from there.

Growing in marshy or wet meadows with other willows, the autumn willow in Colorado is a large shrub, no more than 10 feet tall. The trunks, such as they are, have light gray bark, the branches are light tan, and the ultimate leaf-bearing twigs are a lustrous reddish brown. The contrasting colors of the branches and twigs, the sharp angle at which the branches diverge from the main stems, and the fruits maturing in late summer are some of the distinguishing characteristics of *Salix serissima*.

Yeatts

Neely

Narrowleaf Evening-primrose (Evening-primrose Family)

Oenothera acutissima W.L. Wagner

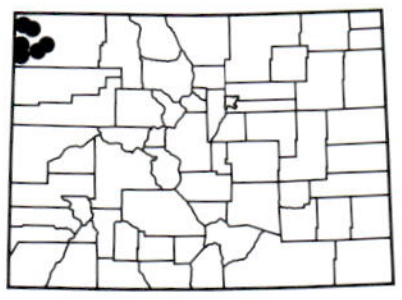

The narrowleaf evening-primrose grows in meadows, depressions, and along arroyos in a variety of habitats ranging from mixed coniferous forest to sagebrush shrublands in northwest Colorado and northeast Utah. The unifying element in these habitats is the presence of seasonally moist, sandy and gravelly soils.

Several similar evening-primroses are found in the area. *Oenothera acutissima* is distinguished by its long, branching taproot, its narrow, irregularly toothed leaves, its large, bright yellow flowers with reddish orange highlights, and its relatively small fruit — a capsule about ¾ inch long.

Evening-primroses are not related to true primroses. Primroses have five-petaled flowers; evening-primroses have either four or two petals. Evening-primrose flowers open in the evening and fade by noon the next day, each flower lasting only one night.

Lesser Club-moss, Northern Spike-moss (Little Club-moss Family)

Selaginella selaginoides (Linnaeus) Link

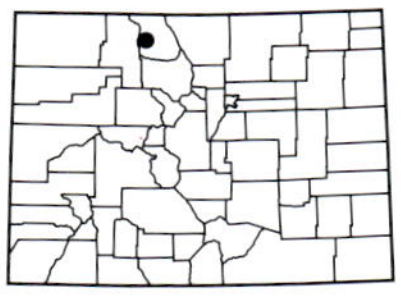

All of the plants described in this book are vascular plants. They contain veins of special cells for transporting water and minerals upward from the soil and food (sugars) away from the leaves where it is synthesized. At one time, all vascular plants were considered members of the division (phylum) Tracheophyta, emphasizing this common trait. More recently, they have been separated into several divisions, which emphasizes the differences among them in other respects. Included in this book are representatives of division Magnoliophyta (flowering plants), division Polypodiophyta (ferns), and division Lycopodiophyta (club-mosses, ground pines, and quillworts). The latter division sometimes is referred to as one of the fern-allies along with other divisions that include the horsetails and whisk-ferns. *Selaginella* is a member of the division Lycopodiophyta.

The various groups of ferns and fern-allies are separated by technical details, the most obvious of which are characteristics of the leaves. The leaves of ferns, conifers, and flowering plants (technically called megaphylls) are conspicuous and have branched veins which leave a gap as the trace(s) to the leaf depart from the central vascular cylinder. Club-mosses, however, have microphylls, leaves with a single vascular trace and no gap.

Despite their appearance, club-mosses are not mosses at all, but are related to the ferns and produce spores. Included in the little club-moss family are the resurrection plants of the deserts of west Texas and Mexico. These are large club-mosses that roll up like a closed fist when dry and then unroll and become greener when the rains return.

Many club-mosses are mossy looking, creeping plants, and they often are found on cool, dry north-facing cliffs or steep canyonsides. However, the lesser club-moss, *Selaginella selaginoides*, is a plant of wet, marshy places such as the mossy banks of beaver ponds or the floor of wet subalpine spruce forests. Although it seldom is seen south of Canada, populations are known from Wyoming. In Colorado, lesser club-moss has been found only once. Dieter Wilken, formerly of Colorado State University, was collecting plants for study by his taxonomy class. He collected some wetland plants in a canyon on the east side of the Park Range and placed them in a terrarium. Later, he discovered he unexpectedly had found the lesser club-moss. Unfortunately, neither Dr. Wilken nor anyone else has managed to relocate the site.

When Linnaeus named the plant in the eighteenth century he believed it to be a *Lycopodium*, a greater club-moss, but he realized it had features of a little club-moss. Later botanists corrected Linnaeus' error, leaving us with a scientific name that translates nonsensically as "selaginella that looks like a selaginella."

NEWFS/Brown

Von Bargen

WEBER MONKEYFLOWER (Figwort Family)

Mimulus gemmiparus W.A. Weber

Steinkamp

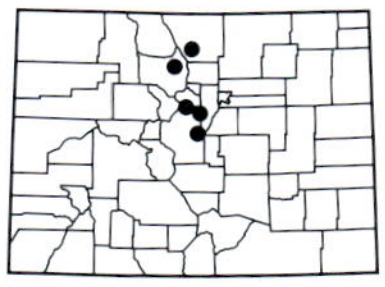

Unique is the only word that describes this plant adequately. First, it is unique because it is an endemic, growing only in a very limited area. Second, it is unique because it does not reproduce in the traditional way.

About 1950, Ruth Ashton Nelson, botanist and author of *Plants of Rocky Mountain National Park*, discovered this monkeyflower on the Grand County side of the park. Since then, the plant has been found elsewhere in the park, as well as in one locality in the Tarryall Mountains. In 1995, it was located near Guanella Pass. *Mimulus gemmiparus* is unknown outside this small area of Colorado. Nelson noticed that flowers were produced infrequently and did not set seed. She experimented with the plant and found that reproduction was carried on by a special structure.

In 1972, William A. Weber of the University of Colorado named and described the plant. The pollen appears to be sterile and cannot fertilize the female reproductive structures. Instead, the petiole (stalk) of the leaf develops into a swollen sac which contains a miniature plant (gemmule). As the mother plant matures, the leaf and plantlet fall and may be carried by water to a new location.

Typically, Weber monkeyflower grows in the moist soil of forest seeps or springs protected by granite overhangs, usually at elevations of 8,500 to 10,500 feet. Amazingly, however, it was found on the alluvial fan deposited by the Lawn Lake Flood in Rocky Mountain National Park, apparently carried down from above by the floodwaters. *Mimulus gemmiparus* often grows in association with other monkeyflowers, *Mimulus rubellus* and *Mimulus guttatus*. It is likely that *Mimulus gemmiparus* is not as rare as we now think but merely overlooked. It is tiny and blends into the wet gravels or tiny crevices in which it grows.

ROUND-LEAVED SUNDEW (Sundew Family)

Drosera rotundifolia Linnaeus

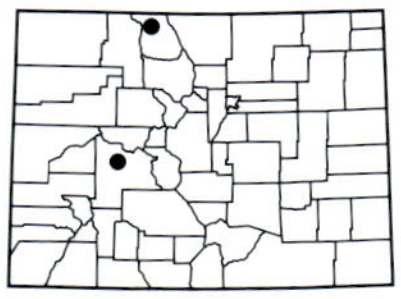

Plants in the genus *Drosera* are carnivorous, capturing insects on the leaves with sticky-tipped, stalked glands that act like miniature tentacles. When the insect becomes stuck on a few glands additional ones then bend inward and enclose the prey. The leaf blade itself also may bend to enfold the captive. The plant then digests the trapped prey, obtaining additional supplies of the essential nutrients nitrogen and sulfur, which are in short supply in the boggy areas in which it grows. Notice that the plant in the photograph has captured a cranefly (lower right portion of photo).

A very short stem bears a rosette of leaves whose blade is broader than long, hence the name *rotundifolia*, meaning round-leaved. Leaf shape is one of the major distinguishing features among species of sundew. The stalked glands, each with a droplet of clear, viscous liquid at the tip, make the plant glisten as if it were covered with dew. The sundew produces several white or pinkish flowers on a leafless stem rising several inches above the leaves.

The species is found across the northern United States, through western Canada and Alaska, and across northern Europe and Asia. Other nearby populations of *Drosera rotundifolia* occur in Yellowstone National Park and Montana. In Colorado, sundew is known from localities in Gunnison and Jackson counties.

Lehr

Jennings

Jennings

Ute Ladies'-tresses Orchid (Orchid Family)

Spiranthes diluvialis Sheviak

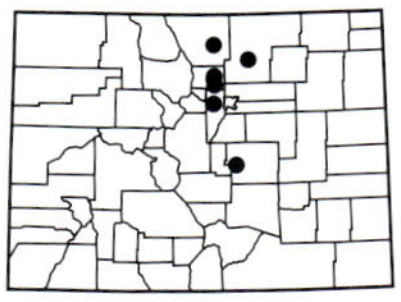

For more than a century, a few specimens of an unusual wild orchid collected in Colorado, Utah, and Nevada lay misidentified in several herbaria. As part of his research on *Spiranthes* orchids, Charles J. Sheviak examined the specimens but could not relate them to known species. In 1980, a small patch of this orchid was discovered near Golden, Colorado, generating a great deal of interest among Colorado botanists. After further field and laboratory study, Dr. Sheviak determined that the plants represented a new species, which he named *Spiranthes diluvialis*.

In Colorado, the Ute ladies'-tresses orchid was known from only three small patches totaling a few hundred plants. But in 1986, as he drove along a freeway, engineer-botanist William F. Jennings spotted what appeared to be an extraordinarily large number of orchids bobbing in the summer breeze. A careful count showed more than 5,500 plants in the group, the largest known concentration of *Spiranthes diluvialis* at the time. In recent years, large populations have been discovered in Utah, and even more recently, small populations have been found in Idaho, Montana, Wyoming, and Nebraska.

All known sites for *Spiranthes diluvialis* are near major streams or along abandoned meanders where ample subsurface water percolates through the stream gravels underlying lush meadows. At some locations, the orchid actually grows in running water.

Ute ladies'-tresses orchids produce a flowering stalk about 14 inches tall arising from a basal rosette of leaves. At the upper few inches of the stalk are numerous small white flowers arranged in a helix like a so-called spiral staircase. The name of the genus, *Spiranthes*, is derived from this feature. *Spiranthes diluvialis* is listed under the Endangered Species Act as Threatened.

Porter Feathergrass, Porter Needlegrass (Grass Family)

Ptilagrostis porteri (Rydberg) W.A. Weber

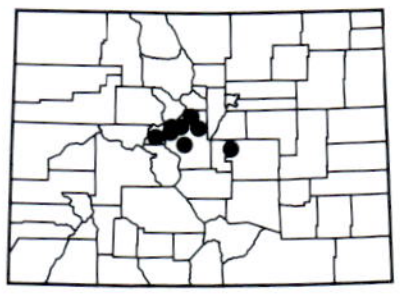

Porter feathergrass was discovered by pioneer botanists Elihu Hall and J.P. Harbour in 1862. In 1874, Thomas C. Porter and J.M. Coulter, authors of *Synopsis of the Flora of Colorado* (Colorado's first flora), called the plant *Stipa mongholica*. In 1905, the plant came to the attention of P.A. Rydberg, a botanist who worked for many years in the West. He decided that the plant did not quite fit the description of the Asiatic *Stipa mongholica*, but that it was a separate species. In recent years, the species related to *Stipa mongholica* all have been transferred to the genus *Ptilagrostis*.

Since the days of Hall and Harbour, *Ptilagrostis porteri* has been reported a few times from the high mountains along the northwest side of South Park and from a few other scattered locations in central Colorado. Porter feathergrass inhabits peat hummocks in willow carrs above 10,000 feet, but only where the valley floor is nearly flat and exposed to the south and east. A special part of the grass flower, called the awn, in *Ptilagrostis porteri* is about an inch long. The awn is bent and twisted and covered with long hairs, giving the appearance that the grass is covered with many tiny feathers. This appearance is the source of the common name, Porter feathergrass. The other common name, needlegrass, often is applied to plants in the genus *Stipa*.

Thellungiella, Salt Watercress, Salt Lick Mustard (Mustard Family)

Thellungiella salsuginea (Pallas) O.E. Schulz

Jennings

The French trappers of the early nineteenth century called South Park *Bayou Salado* in reference to its salt springs and deposits. Two early settlers, John Rollins and Charles Hall, attempted to use the salt springs. In 1864, they invested a reported $50,000 to establish Colorado's first and only salt works, an operation capable of producing 2 tons of salt per day. At that time, salt — and just about everything else — had to be freighted in wagons from the Missouri River at very high cost. However, the railroad was completed from Kansas City to Denver in 1870, and the availability of low-cost transportation effectively put an end to Colorado's great salt industry.

Thellungiella salsuginea is a low, nondescript, annual mustard. Although it is widespread in Asia, there are only a few known locations for the plant in North America. In Colorado, it is limited to South Park. What makes thellungiella interesting is that it grows on the salt flats near Antero Reservoir, within sight of the ruins of the old salt works. Although it is abundant in some years, it is so inconspicuous that diligent searching is necessary to find the first plant. Its white flowers are no help in making the plant stand out against the whitish salt crust.

Thellungiella is a winter annual, meaning that the seeds sprout in late summer and the plant overwinters as a small rosette under the snow. As soon as the snow melts in the spring, the plant bolts, blooms, sets fruit, and dies in a few short weeks, often before anything else blooms in the area. This plant has been very puzzling to botanists, who at various times have included it within the generic concepts of the larger genera *Sisymbrium* or *Arabidopsis*. However, Rocky Mountain botanists William A. Weber and Robert Dorn currently consider the plant to be in the genus *Thellungiella*.

Jennings

Lime-loving Willow (Willow Family)

Salix lanata **Linnaeus ssp.** ***calcicola*** **(Fernald & Wiegand) Hultén**

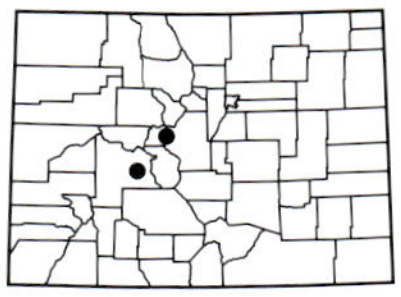

The lime-loving willow, *Salix lanata* ssp. *calcicola*, is known from two locations in Colorado. In 1984, it was discovered by William A. Weber at 12,000 feet in a mountain cirque in Park County. Subsequently, it was discovered in Gunnison County. The other nearest populations are in Banff National Park, Alberta, Canada. This eastern Canadian low-Arctic subspecies is distributed primarily around the shore of Hudson Bay and eastward to northern Labrador. Even the Banff populations are a considerable distance from the main range.

At its locations in Colorado, lime-loving willow forms low, dense, rounded bushes less than 3 feet tall. In Park County, it occurs on loose limestone debris above the edge of a small lake or tarn whose shores are dominated by alpine species. The leaves of *Salix lanata* ssp. *calcicola* are broadly oval and have a whitish, waxy coating on the undersides. The catkins are over 2 inches long and stand stiffly erect.

Colorado Watercress (Mustard Family)

Rorippa coloradensis **Stuckey**

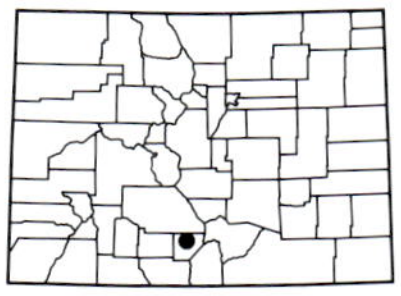

Sometime in 1875, T.S. Brandegee, the botanist with an expedition led by Ferdinand V. Hayden, collected a specimen of watercress, probably at the margin of an alkaline lake and apparently somewhere in the San Luis Valley, although we don't know the exact locality. This specimen, identified as *Nasturtium trachycarpum*, lay undisturbed for nearly a century in the herbarium of the Philadelphia Academy of Sciences until it was examined in 1972 by Ronald L. Stuckey of The Ohio State University. Dr. Stuckey decided that the Brandegee specimen was an undescribed species, which he named *Rorippa coloradensis*.

The Colorado watercress had deeply lobed leaves, with the lobes rounded. At the top of the stem were several white or yellow flowers. The one known specimen was about 6 inches tall but if it was typical of watercresses, it probably was sprawling rather than standing erect. However, this is all we know about the Colorado watercress. Despite several searches, the plant has never been found again. Wetlands in the San Luis Valley have been substantially modified by systems of canals, the digging of artesian wells, and conversion for agriculture. Thus, *Rorippa coloradensis* is presumed extinct.

Martin

Martin

Little Beeplant, Slender Spiderflower (Caper Family)

Cleome multicaulis Sessé & Moçiño ex de Candolle

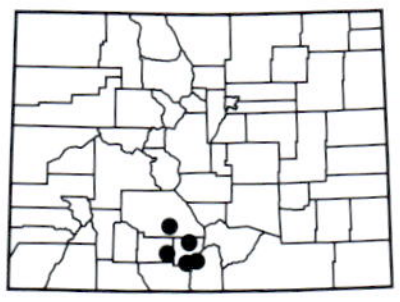

First described in the early nineteenth century from collections in central Mexico, the little beeplant later was found far to the north in the southwestern United States. It is seen around margins of moist, slightly saline depressions, old lakebeds, and greasewood and creosotebush stands. In Colorado, *Cleome multicaulis* occurs in the San Luis Valley, where it typically forms narrow bands about a foot wide in moist soil just outside stands of the rush *Juncus ater*. Unfortunately, these are the areas where grazing animals walk, trampling the little beeplant. The plant apparently is not readily eaten by livestock. In other settings, large clusters or even sizable patches of little beeplant occur, always in very wet, alkaline soils.

Little beeplant is an annual — blooming, producing fruit, and dying in one season. It has thin stems about 30 inches long, allowing it to sway gracefully in the breeze. Its leaves are divided into three leaflets and it has numerous small, delicate, pinkish purple flowers.

First found in Colorado in 1873 by John Wolf, botanist with the Geographical Survey of the Territories, little beeplant is rarely reported. It has been recorded from wetland habitats called *cienegas* in southwestern New Mexico and southeastern Arizona, but recent collections from those areas are few. However, the plant recently was discovered in Wyoming. It appears that little beeplant has disappeared from about 90 percent of its former range, probably because of alteration of wetland habitats. The plant is threatened in the San Luis Valley by the drainage of wetlands and by water tables lowered by pumping for irrigation. Flooding of suitable little beeplant habitat for waterfowl nesting and recreation may also pose a threat. The Nature Conservancy recently acquired a San Luis Valley preserve that will protect one population of little beeplant.

Woodlands

Woodlands differ from forests in that the woodland trees rarely exceed 40 feet in height. Typically, they are open spaced, except in the higher elevations and on moister sites, where crowns may interlock to present a closed, forest-like aspect. The USDA Forest Service takes a practical point of view, defining woodlands as forested land in which valuable timber species make up less than 10 percent of the cover.

The woodlands of Colorado are composed primarily of junipers and pinyon pines. One-seed juniper, *Juniperus monosperma*, and pinyon pine, *Pinus edulis*, dominate on the eastern slope of the Rocky Mountains and extend onto the plains in southeastern Colorado. On the Western Slope and on the Colorado Plateau, pinyon pine shares dominance with Utah juniper, *Juniperus osteosperma*. Rocky Mountain juniper, *Juniperus scopulorum*, occurs in woodlands on both sides of the mountains, usually at the highest elevations of the woodland zone. In Colorado, some 6 million acres — about 10 percent of the state's area — may be considered woodlands. Most woodlands occur in the western third of the state, but sizable areas also exist on the Eastern Slope, particularly south of Colorado Springs. About 25 percent of the area considered woodlands is covered with scrub oak (*Quercus* spp.), and 16 percent is composed of pure juniper stands.

Evidence indicates that woodlands were displaced far to the south during the Pleistocene glaciations, becoming established in our region afterward, about 10,000 years ago. It appears that the woodlands have expanded in the last 2,000 years, perhaps in response to climatic change. The ecology of natural disturbances in Colorado woodlands is not well known. Although grazing animals always existed in woodland areas, the woodlands did not evolve in response to heavy grazing, as did the plains. Fires certainly occurred, but they appear to have been very infrequent, occurring only every few hundred years. Fire frequency depends upon the amount of fuel available and the degree of closure of the woodland. If the canopy is open, grass can grow between the trees and fuel ground fires that may scar the trees, but probably would not kill them. If the canopy is closed, grass is limited and fire would be even less frequent. When fire occurs, however, it probably would kill the trees as a crown fire. In general, the low rate of natural disturbance in most woodlands has allowed these trees to reach great ages — 700 to 1,000 years in some areas.

Woodlands occur in diverse geologic settings, as underlying formations can consist of igneous, metamorphic, or sedimentary rocks. Soils often are poorly developed and quite dry. Woodlands can grow even where there is little soil, just cracks in the bedrock. This rich variation in geology creates many opportunities for specialized plants, and many rare plants occur on only a single geologic formation. Other rare plants such as Grand Junction milkvetch, *Astragalus linifolius*, or Payson lupine, *Lupinus crassus*, occur on several similar formations, but only where they are exposed in the same manner.

The woodland habitat was used extensively by early peoples. The cliff dwellings of southwestern Colorado, including those at Mesa Verde, are located within the woodlands. Wooded mesa tops were cleared for farming and woodland trees were cut for firewood. Later, the Ute Indians used the area, consuming the edible pinyon nuts as an important part of their diet. Unlike woodlands in Utah and Nevada, most of Colorado's woodlands escaped use for charcoal production. However, our woodlands were heavily grazed, particularly from 1880 to 1920. Recently, pinyon firewood has become very popular and use of woodlands as a firewood source is increasing. Also, some areas have been cleared for farming, especially for pinto bean production in the southwest corner of the state. Other areas have been "chained" to increase grass production for livestock grazing. Chaining is a destructive process in which a heavy chain, attached between two bulldozers driven parallel to one another, is dragged through the woods, knocking down or uprooting all trees in its path.

A few areas of woodland, including those in national parks or monuments, are effectively preserved. Some other sites have been set aside as natural areas. But the rich diversity and unusual plants present in woodlands indicates a need for additional preservation of this habitat.

Martin

Paradox Valley Lupine, Payson Lupine (Pea Family)

Lupinus crassus Payson

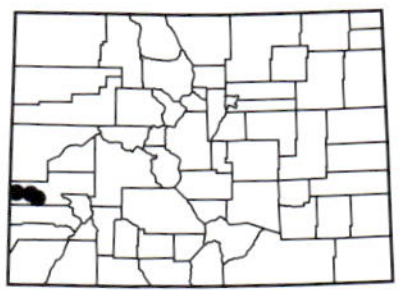

The Paradox Valley lupine is known only from western Montrose County in the vicinity of Paradox Valley, where it was discovered in 1914. It grows beneath junipers on fairly open ground, but also may be found in stands of mixed pinyon and juniper. It usually grows on sandy soils derived from the Dakota, Burro Canyon, and Chinle formations, but there also are populations on adobe hills. Paradox Valley lupines are large perennial plants whose stem and thick leaves tend to sprawl with only the flowering stalks held erect. The plants have white flowers with purple tips, and they bloom in May.

The widespread genus *Lupinus* contains many plants of horticultural value such as bluebonnet, the state flower of Texas. On the other hand, many lupines are poisonous, particularly to sheep.

Field observation suggests that the seed crop of *Lupinus crassus* suffers heavy predation by insect larvae. Mining and road construction are possible threats to the habitat of this species.

Pagosa Phlox (Phlox Family)

Phlox caryophylla Wherry

The attractive Pagosa phlox is limited to La Plata and Archuleta counties in Colorado and adjacent Rio Arriba and Taos counties in New Mexico. The plant's range is generally southeast and southwest of the town of Pagosa Springs, Colorado, from which the common name is derived. Pagosa phlox, *Phlox caryophylla*, occurs from about 6,500 to 7,500 feet in elevation on open grassy slopes or in open woods, frequently near scrub oak. Pagosa phlox is an erect plant, about 8 inches tall and arising from a rather woody base. Its leaves are very narrow and linear, about 1/8 inch wide and 2 inches long. The flowers, appearing mainly in June, are a striking pink or purplish pink and are almost an inch across.

The entire U.S. Highway 160 corridor from Pagosa Springs to Cortez is a highly popular region with resorts and summer homes dotting the area where only open stands of ponderosa pine and oak brush once existed. Habitat for *Phlox caryophylla* and other rare plants has been diminished and will continue to decline as development continues. However, Pagosa phlox appears to tolerate some disturbance to its habitat, and there are populations known by the roadside.

Jennings

Jennings

Steinkamp

Ripley Milkvetch (Pea Family)

Astragalus ripleyi Barneby

With its pendulous, pale lemon yellow flowers, Ripley milkvetch is a tall, striking plant that brightens canyons and mesa top woodlands at the southwestern perimeter of the San Luis Valley. Following the flowering period from late June to late July, striking, pendent pods remain. Limited to an area straddling the Colorado-New Mexico state line in the Rio Grande and Conejos River valleys, *Astragalus ripleyi* is found in grassy ponderosa pine or aspen savannas and in sagebrush and rabbitbrush flats adjacent to watercourses. It grows mainly in soils derived from the volcanic rocks of the San Juan Mountains. Plants occasionally are seen in the open in areas where livestock grazing has been light. In heavily grazed areas, most plants are found growing up through shrubs. Grazing may be limiting the plants on these sites, but it also is possible that they require shade from a nurse plant during seed germination and seedling establishment.

Martin

When the plant was collected in 1950 near Tres Piedras, New Mexico, Rupert Barneby realized it was a new species. He named it for his friend and collecting partner, H.D.D. Ripley. Later, William A. Weber found the plant in Colorado in an area where it had been collected in 1935 by Francis Ramaley. Professor Ramaley was the first botanist to record the plant, but apparently he did not realize it was an undescribed species. In 1986, Steve O'Kane, Jr., and John Anderson made an intensive search for Ripley milkvetch and located several new populations.

Steinkamp

The most significant threats to Ripley milkvetch appear to be heavy grazing and fire suppression, which allows the forest canopy to fill in and shade out this plant of the open woodlands. *Astragalus ripleyi* is receiving management attention from public land agencies.

Woodside Wild Buckwheat (Buckwheat Family)

Eriogonum tumulosum (Barneby) Reveal

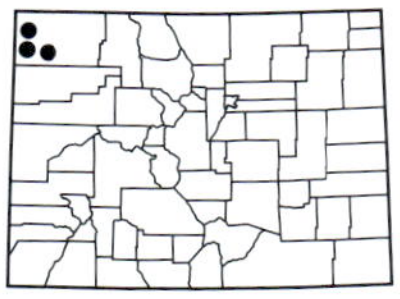

Woodside wild buckwheat, *Eriogonum tumulosum*, is a low mat-forming plant with white flowering heads that barely reach above the leaves. The buckwheat was named for its occurrence near the tiny village of Woodside, Utah. The species' main range is in northeastern Utah, mostly in the Uinta Basin, with a few populations in Moffat County, Colorado, near the Utah state line. Prior to intensive botanical surveys in Dinosaur National Monument there was only one known population in the monument, near Gates of Lodore. Four new Colorado locations were discovered during the surveys of 1987 to 1989.

Jennings

Three of the sites in the monument are more or less inaccessible and seem to be undisturbed. Two others have been grazed by livestock. This gives an opportunity for long-term monitoring of sites with different land use to determine the effect of each use on this rare plant. Simply finding a site and observing it one time does not tell us whether the plant is holding its own, expanding, or declining — important facts to know if land use managers are to make informed decisions. This takes commitment, time, and funding, all of which can be difficult to obtain. If we are committed to the bald eagle, black-footed ferret, and California condor, we also must be committed to their habitats and prey, and to their prey's habitats and food. All elements of the natural world are important.

Jennings

Jennings

SPINELESS CLARET CUP CACTUS, SPINELESS HEDGEHOG CACTUS (Cactus Family)

Echinocereus triglochidiatus Engelmann forma ***inermis*** (K. Schumann) Rowley

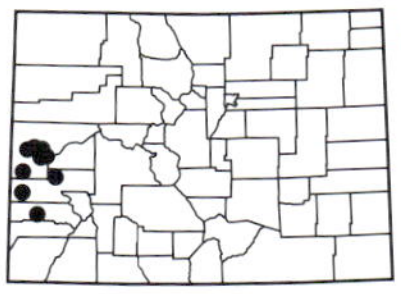

Claret cup cactus, *Echinocereus triglochidiatus*, is a very common but highly variable cactus that occurs over much of the Southwest. In western Colorado and eastern Utah there is a most unusual spineless form of this attractive cactus. The form name, *inermis*, meaning unarmed, reflects this spineless characteristic. In most other aspects, the spineless version of the cactus looks very much like its spine-covered relatives. Typically growing in a cluster of five to fifteen stems, each ribbed and cylindrical stem is about 4 inches tall and 2 inches wide. In May, the plants bear spectacular 3-inch-wide scarlet flowers which are visible from many feet away. Although it was discovered in the late nineteenth century, the spineless form was unreported until its rediscovery in the early 1970s during thesis research by Gerald Arp, then a student at the University of Colorado.

The beautiful flowers and lack of spines cause *Echinocereus triglochidiatus* forma *inermis* to be much sought after by gardeners. This cactus was listed in 1979 as Endangered under the Endangered Species Act. Because removal, transport, or sale of endangered plants is prohibited, there was at least theoretical protection from collectors. However, cactus taxonomy is controversial, and there was a wide variety of opinion among specialists about the taxon, which led to its being removed from the list in 1993. Nonetheless, it is a special plant form that should never be collected but left for all to enjoy.

Grand Junction Milkvetch (Pea Family)

Astragalus linifolius Osterhout

Jennings

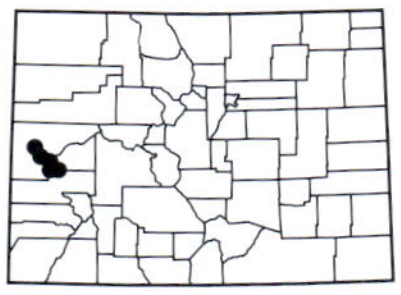

After finding this milkvetch ". . . 6 miles across the Colorado River from Grand Junction," George Osterhout described *Astragalus linifolius* in 1928. More than 50 years passed before the next specimens were collected. Today, the plant is known from a few canyonsides in the pinyon-juniper country of Mesa, Delta, and Montrose counties. There are many closely related endemic species of *Astragalus* on the Colorado Plateau, and the relationship of *A. linifolius* to similar species still is uncertain.

Wilken

The red sandstone cliffs at Colorado National Monument are formed mostly from Wingate Sandstone. Sandy clay soils or shales derived from the Wingate Sandstone and the Morrison and Chinle formations are prime habitat for the Grand Junction milkvetch. The plant may appear leafless as its leaves have only a few small, linear leaflets. In contrast, *Astragalus linifolius* in bloom is all the more astonishing. It produces an attractive display of hundreds of white flowers which have a purple spot on the keel and two fused petals enclosing the stamens. The erect red pods are nearly as attractive as the flowers.

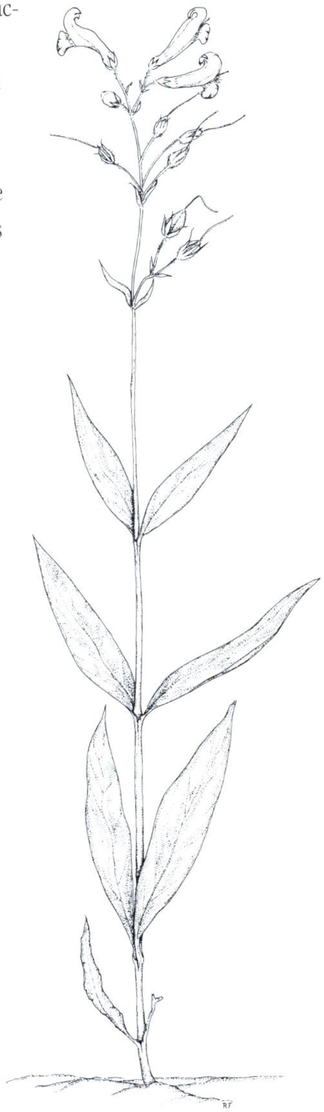

Degener Beardtongue, Degener Penstemon (Figwort Family)

Penstemon degeneri F. Crosswhite

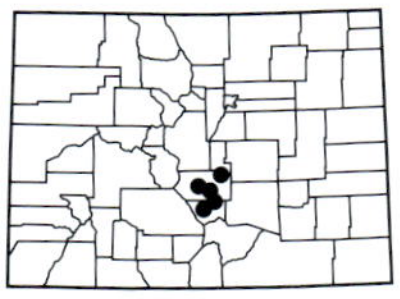

The Degener beardtongue, *Penstemon degeneri*, is found in pinyon-juniper woodlands of the Arkansas River Canyon and in a few other areas near Cañon City. This is one of the least-known penstemons in Colorado, although it can be locally common on decomposing granitic soils south of the Arkansas River. Frank Crosswhite, a specialist in the genus *Penstemon*, saw a specimen obtained by Otto Degener and in 1965 named the plant for Dr. Degener.

Penstemon degeneri is a foot or more tall and has very leafy stems. Toward the end of the stems are several dark blue flowers, each a little less than an inch long. Deep inside the flower, the stamens (pollen-producing organs) have no hairs, but the staminode, a non-functional stamen, is bearded with golden yellow hairs. This feature of many penstemons gives rise to their common name, beardtongue.

Steinkamp

Martin

ARKANSAS CANYON STICKLEAF (Loasa Family)

Mentzelia densa Greene

In the rugged terrain along the Arkansas River between Salida and Cañon City, where mountain mahogany and Gambel oaks cling precariously to the cliffs, the Arkansas Canyon stickleaf is found. The perennial plants form a hemispherical mound up to a foot high and in mid-summer produce yellow sunburst-like flowers. In contrast to the many rare shrubby plants found on soils derived from shale, *Mentzelia densa* is found on granite.

The common name, stickleaf, is applied to many plants in the genus *Mentzelia* because hairs on the leaf cause the plant to stick to clothing. These peculiar hairs are shaped like miniature Chinese pagodas. When the leaves are pulled off clothing, there is a little "r-r-r-rip" sound as the hairs release the fabric. The fruit, an oblong capsule about ½ inch long, contains more than fifty small winged seeds. The capsule is believed to be dispersed by sticking to animals' fur and falling off elsewhere.

Mentzelia densa was collected by a number of botanists from the 1870s to the 1920s, but it was rarely documented during the succeeding years. In 1989, Janet Coles of the Colorado Natural Areas Program, a state agency with the responsibility of assessing natural areas that include habitats for rare plants and animals, intensively surveyed the Arkansas River Canyon for the stickleaf. Her work revealed that in the narrow area between Parkdale and Cotopaxi, the plant is relatively common in disturbed areas on roadsides, railroad embankments, and sandy washes where soils are bare. Thus, the Arkansas Canyon stickleaf seems to typify the kind of rarity in which the plant is locally common within an extremely limited range.

Arkansas Canyon stickleaf's steeply sloping habitat of erosion-resistant granite has protected it from such threats as trampling or grazing by livestock. However, major road relocations along U.S. Highway 50 could seriously affect known populations.

CANADIAN RIVER SPINY ASTER (Daisy Family)

Herrickia horrida Wooton & Standley

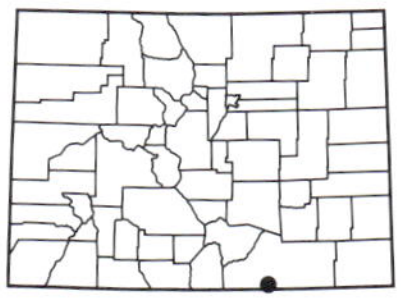

In 1897, C.L. Herrick, a geologist, plant collector, and president of the University of New Mexico, found an unusual aster amidst the oak brush on the hillsides above Ratón, New Mexico. His collection came to the attention of E.O. Wooton and Paul Standley, who were preparing a flora of New Mexico. They not only recognized the species as new, but decided that it did not belong in the genus *Aster*. In 1913, they placed it in the new genus *Herrickia*. Since that time, botanists have argued whether Canadian River spiny aster is sufficiently unusual to be placed in its own genus, and some prefer to call it *Aster horridus*. In either case, it is a unique species.

Steinkamp

Despite being known for 90 years, *Herrickia horrida* has been found only in three northern New Mexico counties and Las Animas County, Colorado. It grows in dry sandy or rocky soils, sometimes on almost bare sandstone bedrock. It apparently is able to tolerate both the baking heat of summer on south-facing rocky outcrops and the cold of winter in the Ratón Pass area. Canadian River spiny aster is easy to identify. It has been described as an "aster on a *Mahonia*," as if someone grafted a typical purple late-summer aster onto a hollygrape, *Mahonia repens*.

SCHMOLL MILKVETCH (Pea Family)

Astragalus schmolliae C.L. Porter

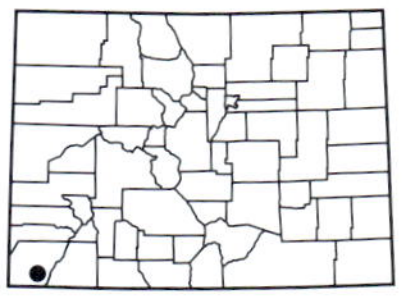

Schmoll milkvetch grows on sandy, gravelly mesa tops and terraces in the pinyon-juniper woodlands of a small part of Mesa Verde National Park. Closely monitored by National Park Service personnel in recent years, this milkvetch was noted to have become the dominant plant in an area disturbed by pipeline construction. *Astragalus schmolliae* seems to favor a thin, red, wind-deposited (aeolian) soil. Although it grows at the end of one of the mesas, the plant apparently has been unable to cross canyons and establish itself in similar soil on other mesa tops.

First collected at Mesa Verde by Alice Eastwood in 1890, *Astragalus schmolliae* was not reported again until 1925. The species was not described until 1945, when it was named for Hazel Schmoll, one of the 1925 collectors.

The Southwest has numerous endemic, localized species of *Astragalus* which differ only subtly from close relatives. *Astragalus schmolliae* is most similar to the widespread species *Astragalus lonchocarpus* and *Astragalus scopulorum*. The flowers of Schmoll milkvetch are yellowish white or creamy and have black hairs on the calyx. The leathery pods, similar to many others in the pea family, curve downward.

Martin

Steinkamp

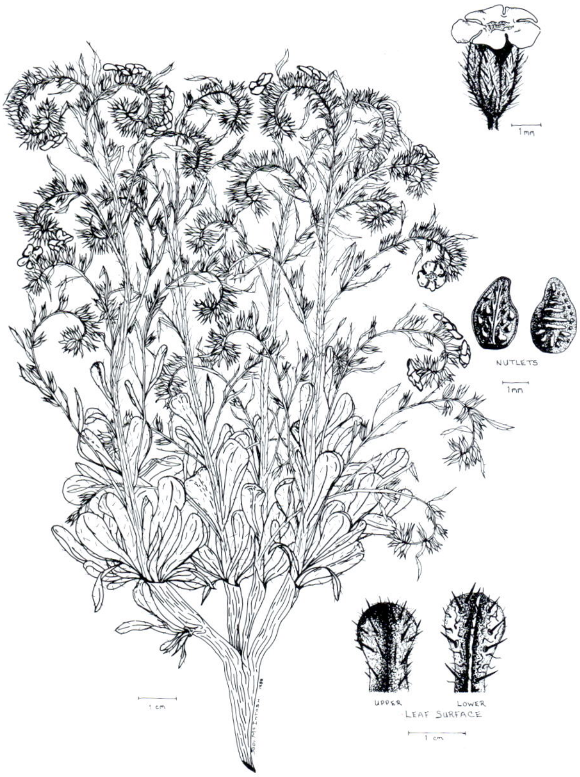

GRAND JUNCTION CAT'S EYE (Borage Family)

Cryptantha aperta (Eastwood) Payson

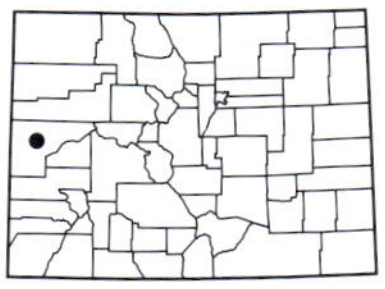

On two occasions in May and June of 1892, Alice Eastwood found an unusual plant growing in the Grand Junction area. It has never been found since. Many botanists have examined her specimens over the years, the majority declaring them to be distinct, judging from the flowers and nutlets. In spite of intensive search, particularly by Walt Kelley of Mesa State College, *Cryptantha aperta* has not been rediscovered. So, it is presumed to be extinct.

Alice Eastwood came to Colorado as a teenager. She later taught high school in Denver. Eastwood collected the cat's-eye in 1892, just before she became curator of botany at the California Academy of Sciences in San Francisco. She was active in that position until 1950, when she retired at age 90. She discovered and described numerous species from all over the West, but locality data on her specimens often are very general. It is presumed that she simply listed the town in which she found herself at the end of the day. Her original field notes might have given the specific location, but they were destroyed along with the California Academy building and most of its contents in the fire that followed the 1906 San Francisco Earthquake. However, Alice Eastwood's heroic rescue efforts before the building was consumed by fire saved 1,497 irreplaceable type specimens that are of inestimable value to science.

Steinkamp

Jennings

Dolores Skeleton-plant (Daisy Family)

Lygodesmia doloresensis S. Tomb

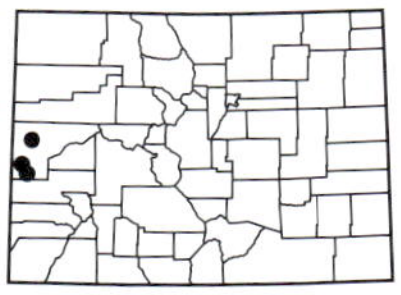

The Dolores River cuts a nearly vertical canyon through the red sandstones of southwestern Colorado. Although there is a good road at river level through some of the canyon, this area still is remote and seldom visited. A number of rare plant species have been found here in recent years as a result of renewed interest in the botany of Colorado's Western Slope. In 1947 and 1949, specimens of the Dolores skeleton-plant were collected by H.D. Harrington of Colorado State University. As part of his doctoral research on *Lygodesmia*, Spencer Tomb examined Harrington's specimens, decided they represented an unrecognized species, and described *Lygodesmia doloresensis* in 1980.

There is one extended population of Dolores skeleton-plant along 10 miles of the canyon, growing between the canyon wall and the river in reddish purple, sandy soil derived from the Cutler Formation. A population was found in 1988 along the Colorado River in Utah and another is located in Rabbit Valley west of Grand Junction. Domestic livestock grazing has severely affected this species, and plants often are found only where they are protected by shrubs or clumps of prickly pear cactus. If livestock pressure is low, plants frequently are found on sandy outwash slopes at the bases of cliffs.

Standing about a foot tall, the Dolores skeleton-plant produces many flowering stems that branch near the base of the plant. Leaves are narrow, less than ⅛ inch wide, but up to 5 inches long. Blooming in June, *Lygodesmia doloresensis* produces numerous flowers rose to lavender in color.

Knowlton Cactus (Cactus Family)

Pediocactus knowltonii L. Benson

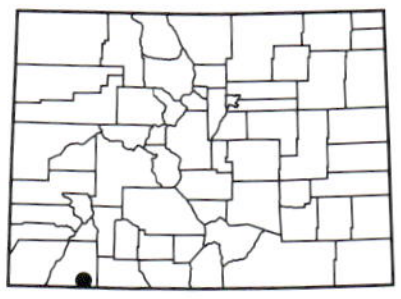

On a dry, gravelly hilltop covered with pinyon pine and juniper near the Colorado-New Mexico state line are a few hundred plants of this tiny, extremely rare cactus. When *Pediocactus knowltonii* was discovered in 1958 by Fred Knowlton, who was building a road, the population was estimated to be in the thousands. After the species became known to the public, cactus poachers decimated the population, almost exterminating it. When the San Juan River was dammed to create Navajo Lake, poorly informed would-be rescuers, believing the population was to be inundated, further depleted the colony. The owner of the property donated the land to The Nature Conservancy, which has erected a strong fence around the cactus patch to discourage cactus rustlers and prevent trampling by cattle. In 1979, this species was afforded federal protection when it was listed as Endangered under the Endangered Species Act. Removal, transport, or sale of endangered plants is prohibited.

Martin

Knowlton cactus is an extremely small, inconspicuous plant. Usually, it is less than an inch wide and barely protrudes above ground level. Unless the plant is blooming, it is virtually impossible to see, even when one is standing next to it. The pinkish flowers, which appear at the end of April, are nearly as large as the cactus itself.

MESA VERDE STICKSEED (Borage Family)
Hackelia gracilenta (Eastwood) Johnston

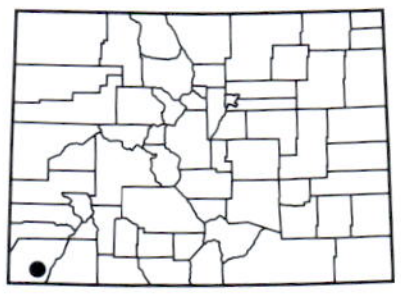

In the dense litter beneath oaks or pinyon pine and juniper in Mesa Verde National Park are a few plants of Mesa Verde stickseed, *Hackelia gracilenta*. Fortunately, this rare species is monitored closely by the National Park Service. First believed limited to moist canyon sites, *Hackelia gracilenta* has been found in recent years on the mesa top, as well. Thousands of plants appeared in an area where the pinyon-juniper woodland burned in the late 1980s. The plant clearly is adapted to a landscape where fire is a normal event. Our tendency to suppress fire may have made Mesa Verde stickseed less abundant now than it was prehistorically. A major fire that occurred in the park during the summer of 1996 may provide habitat for another explosion in the stickseed's population.

Jennings

Steinkamp

National Park Service research also suggests that the expansion of the plant's range is related to above-average precipitation that occurred at Mesa Verde during the 1980s. However, the added moisture in the canyons has caused the grasses to grow more luxuriantly there. As a result, they have out-competed the stickseed in its usual habitat. If drier years return, will the stickseed die out on the mesa tops? If the high grasses in the canyons have reduced those populations, will there be enough canyon plants to maintain the species in dry years? Nobody knows.

The stems of stickseed are clothed with stiff, bristle-like hairs. The nutlets bear hooked structures that attach themselves to clothes or animal fur, giving rise to the common name. Leaves mostly are crowded toward the base of the plant, and the funnel-shaped flowers are blue or whitish blue.

Steinkamp

PARK ROCKCRESS (Mustard Family)
Arabis vivariensis S.L. Welsh

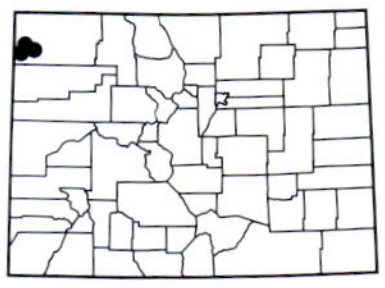

Beginning in the mid-1950s, Stanley L. Welsh of Brigham Young University occasionally encountered a rockcress from the Utah portion of Dinosaur National Monument. Noting that the plants were "peculiar," he sent specimens to a mustard family specialist who identified the plants first as *Arabis microphylla* and upon reexamination 20 years later as *A. microphylla* with some features of *A. fernaldiana*. Dr. Welsh recognized that his collections showed many similarities to the latter species but felt the differences were significant enough that the plants from the monument represented a distinct species, which he described in 1986 under the new name *Arabis vivariensis*. Furthermore, *Arabis fernaldiana* is unknown east of the Elko, Nevada, area. The entire width of the state of Utah separates the Nevada plants from the Dinosaur National Monument rockcress populations.

Beginning in 1987, botanists Steve O'Kane, Jr., Walt Kelley, Tamara Naumann, and Betsy Neely intensively investigated the flora of Dinosaur National Monument. At that time, only nine specimens of *Arabis vivariensis* were known. In the ensuing three field seasons, twenty-one populations were identified. Most colonies were found on shady north-facing slopes or in the shade of pinyon-juniper woodlands. The park rockcress is usually associated with sandy soils derived from the Weber Sandstone. A few populations have been identified in Uintah County, Utah, but its center of distribution appears to be Dinosaur National Monument.

Park rockcress forms mats or carpets sometimes more than 3 feet wide. The plant spreads horizontally, but the ends of the branches turn upward to bear the flowering stems or leaf rosettes of the current season. The delicate flowering stems can be up to a foot tall and are covered with minute, branched hairs. The flowers are white, fading to soft pink, but they are seldom seen because the plant blooms in April and is in fruit by mid-May. The fruits, or siliques, are a striking feature of the plant, as they are thin, whiplike, and up to 6 inches long.

Suggestions for Further Reading

Ehrenfeld, D. (ed.). 1995. *Plant Conservation: Readings from Conservation Biology*. Blackwell Science, Inc., Cambridge MA.

Leopold, A. 1949. *A Sand County Almanac*. Oxford University Press.

Noss, R.F. and A.Y. Cooperrider. 1994. *Saving Nature's Legacy. Protecting and Restoring Biodiversity.* Island Press, Washington, D.C.

Spackman, S., W.F. Jennings, J. Coles, C. Dawson, M. Minton, A. Kratz, C. Spurrier, and T. Skadeland. 1997. *Colorado Rare Plant Field Guide*. Prepared for the Bureau of Land Management, the USDA Forest Service, and the U.S. Fish and Wildlife Service by the Colorado Natural Heritage Program.

Weber, W.A. and R.C. Wittmann. 1996. *Colorado Flora: Eastern Slope*. Revised edition. University Press of Colorado, Niwot, Colo.

Weber, W.A. and R.C. Wittmann. 1996. *Colorado Flora: Western Slope*. Revised edition. University Press of Colorado, Niwot, Colo.

Wilson, E.O. 1992. *The Diversity of Life*. Belknap Press of Harvard University Press, Cambridge, Mass.

Many other pertinent references can be found at the Denver Botanic Gardens library, the best Colorado botanical library open to the public.

Index

Index

Index

"...the loss of genetic and species diversity by the destruction of natural habitats...is the folly our descendants are least likely to forgive us."

— Dr. E.O. Wilson

This rare plant habitat in the San Luis Valley no longer exists as pictured.

Kuntz